Praise for *Scientifi*

This little book is a must-read for anyone who is passionate about their research; who wants to communicate complex information effectively; who cares about securing grant funding or investment, or who cares about advancing their research career. That is to say it is essential reading for *everyone* involved in science or scientific communication, including researchers, company executives, doctors and policy makers. In a world where there is increasing competition for an audience's attention, this book provides a short, easy-to-read, practical guide to ensuring your messages win hearts and minds. Written from a place of empathy for those who are reticent about public speaking and engagement, but with plenty to offer veterans too, it practises exactly what it preaches: it is engaging, direct and highly effective. I cannot recommend it highly enough.

Vivek Muthu
Physician and health and life sciences consultant
Former Managing Director, Economist Intelligence Unit Healthcare

I had the great pleasure to train with Jo Filshie Browning before, and this experience has been truly transformative. In her book, Jo distills her extensive knowledge to help scientists navigate the complex landscape of scientific communication. Combined with illustrative anecdotes and real life stories, the lessons on communication theory and how proper communication can work for you are elegantly interwoven, and thus make for an easy but very informative read. I really enjoyed reading the book and highly recommend it to all scientists who want to make their science more impactful by taking the next steps communicating it effectively.

Benedikt Westphalen
MD Medical Lead Precision Oncology, University of Munich

Packed full of helpful tips and guidance, examples that resonate and statements that stick, Jo Filshie Browning has created a guide that is informative, comprehensive, easy to follow and immensely practical. If your goal is to communicate your scientific data as powerfully as possible, then *Scientifically Speaking* is the book for you. Jo's words have such a strong call to action. I was totally inspired to get out there and connect with my audience through compelling presentations. As Jo says, "science doesn't speak for itself". With the help of this book, you will certainly be able to provide it a mighty and extremely impactful voice!

Sangeeta Jethwa, MD
Research Physician
Chief Medical Officer, Elaaj Bio AG

As a physician, it has become increasingly important to not only spread scientific education by digital means and social media, but also fight the vast amount of disinformation that is being spread. In order to do this, you need the right tools and training. This is not just a textbook, but a great reference to guide you as you deepen your engagements and communication in this brave new digital world. The material is discussed in a clear, concise and practical way, that makes it perfect for the working physician who wants to form part of the digital conversation.

Gilberto Morgan MD
Medical oncologist
Director of the OncoAlert network

If you have ever felt even slightly challenged with preparations to deliver your medical or scientific message effectively, Jo Filshie Browning clears the way by providing very smart, practical and effective advice on how to address different audiences and in a variety of settings. Add to that numerous compelling anecdotes from her 25 years in high-level healthcare communication and very timely examples from the COVID-19 crisis and you have a book you literally do not want to put down. This book will teach you not only how to prepare your content in the most thorough manner, but also how to successfully train your voice, master your body language and appearance to best support your communication goals.

It is an essential read and powerful toolkit for anyone in the medical and scientific professions, and for everyone interested in communicating in the most impactful way.

Gabriele Y. Matthias
Molecular Biologist, Friedrich Miescher
Institute for Biomedical Research

As scientists we might have important ideas and research about how to better the future of humankind, but often we do not know how best to communicate our concepts so that they can be fully and easily understood. The only way to ensure that science can have impact is to narrate our science clearly and in a compelling way. This is where Jo Filshie Browning's book *Scientifically Speaking* has an important role to play. I know personally that principles in this book, when scientifically applied, can lead to the best possible impact. I worked with Jo on my TEDx talk and her practical and clear principles of scientific communication were extremely valuable in my preparation and the talk was a huge success and was featured on the main TED platform. I do hope others will read this and benefit from her approach.

Antonella Santuccione Chadha
Co-Founder and CEO of Women's Brain Project,
Euresearch Vice-President
Head of Stakeholders Engagement Biogen

Being a presenter and media interviewee on scientific/medical topics, Jo Filshie Browning's book rang many bells for me, while providing fresh insights and practical advice. Presentation strategies here are both ancient (Aristotle's *ethos – pathos – logos*) and very modern, in this age of virtual webinars and conferences, social media echo chambers, ultra-short attention spans, science denialism and fake news. For me the top 'take-home' message was *'prepare, prepare'* supremely for the questions after a presentation! Browning uses the classic 1995 "Pill-scare" to exemplify how better media planning might have reduced unintended conceptions generated by scary headlines, about doubled *relative* risk of thrombosis, rather than the equally valid *absolute* risk increasing from 1:7000 to 2:7000. This great read will minimise similar 'lost in translation' scenarios between science fact and public understanding.

John Guillebaud
Emeritus Professor of Family Planning and Reproductive
Health, University College London

The book *Scientifically Speaking* by Jo Filshie Browning is a must-read for all scientists wishing to enhance the impact of their research and further their career. The author breaks down the many communication complexities providing a clear strategy and various tools and methods. This book is a go-to reference to tackle our fast-changing world of communication technologies.

Maarten Vercruysse, PhD
Principal Scientist
Roche Pharma Research & Early Development

Scientifically Speaking provides clear, fundamental guidance for scientists and physicians regarding effective audience-appropriate and media-appropriate communication of scientific and clinical findings. The book should be especially helpful for scientists and physicians who do not have access to communications professionals.

Joseph Bolen, PhD
Chief Scientific Officer
PureTech Health

Scientifically Speaking

HOW TO SPEAK ABOUT YOUR RESEARCH WITH CONFIDENCE AND CLARITY

JO FILSHIE BROWNING

First published in Great Britain by Practical Inspiration Publishing, 2021

© Jo Filshie Browning, 2021

The moral rights of the author have been asserted.

ISBN   9781788602785 (print)
        9781788602778 (epub)
        9781788602761 (mobi)

All rights reserved. This book, or any portion thereof, may not be reproduced without the express written permission of the author.

Every effort has been made to trace copyright holders and to obtain their permission for the use of copyright material. The publisher apologizes for any errors or omissions and would be grateful if notified of any corrections that should be incorporated in future reprints or editions of this book.

Practical Inspiration Publishing

FSC
www.fsc.org
MIX
Paper from responsible sources
FSC® C013604

For my parents, Professor Marcus Filshie and Dr Sheila Filshie, the best science communicators I know.

Contents

List of figures .... *xi*
Foreword .... *xiii*
Speaking about science in the digital age 1

Part 1: Why you need to speak, and how it works 5
    Chapter 1 The speaking advantage 7
    Chapter 2 Setting the scene for speaking... 19

Part 2: The principles of persuasive speaking and presenting 31
    Chapter 3 What to say... 33
    Chapter 4 How to say it 45
    Chapter 5 Mastering the Q&A 65
    Chapter 6 Communicating numbers, risk and uncertainty.................. 81

Part 3: How to speak engagingly in any type of situation................. 95
    Chapter 7 Energy.. 97
    Chapter 8 Managing your nerves and speaking with
        confidence 103
    Chapter 9 Speaking in different settings: Virtual meetings and
        other formats 115
    Chapter 10 Winning the war of attention... 135
    Chapter 11 Physical and vocal exercises for
        speaking success 137

The author *147*
Acknowledgements *149*

List of figures

1 The transmission model for communication 20
2 The media approach to storytelling vs the scientific and academic approach 27
3 Interpreting information can often depend on your perspective 34
4 Start by asking 'With what effect?' 35
5 Establish who you're speaking to 35
6 Next focus on your main message 38
7 Your messages in the form of a Greek temple 41
8 The Q&A – your topics 67
9 How you come across as a communicator is critical in successful communication 98
10 It's important to understand the characteristics of each channel 115
11 It's important to ensure that you light your face well from behind the camera 120
12 Ensure your camera level is in line with your eye line 122
13 A reminder to maintain your eye contact with the camera 123

Foreword

During 2020 the world waged war on a pandemic. Doctors and scientists were frontline in this war, treating, innovating and developing treatments to reduce the huge human impact of COVID-19. As a long-time scientist, there is however another war that is raging that we need to understand – the war on knowledge and reason itself.

How the world finds and consumes information has been transformed over the past 20 years. Amidst online technology superpowers like Facebook and Google, the collapse of traditional media and a click-bait fueled Internet, the conditions were ripe to cause an information tsunami, exploiting the fear and anxiety during the pandemic. Anyone could become an expert – sharing essays and tweets on the epidemiology or transmission of the virus that were completely false. No fact-checking, no verification – these tweets became weaponised.

Another example was in Iran, when during the early days of the outbreak, a message was spread that drinking bootleg liquor was effective at keeping the virus at bay. Amidst the fear, tens of thousands took the message as truthful, thousands were hospitalised and over five hundred people died due to this hoax, spread so easily by technology. The World Health Organisation called it an 'infodemic' and took the extraordinary step of calling all scientists to help "to promote access to health information and to mitigate harm from health misinformation among online and offline communities".

Nothing is more important than for science (and therefore scientists) to stand up, communicate and give facts and falsehoods a voice in the midst of this knowledge war. This is why all scientists must read *Scientifically Speaking* to better equip themselves on how to communicate. Let me explain why.

The pandemic has fundamentally altered how people engage, learn and value science. The good news is that the world now values science more than ever. The breathtaking speed in developing effective COVID-19 vaccines has made people appreciate the scientific enterprise. But just as innovation has accelerated in part due to technology, so has the volume and speed of lies, misinformation and falsehoods. Technology platforms incentivise and weaponise the spread of conspiracy theories, reaching far more people than previous generations. One recent MIT-led study found that lies spread six-times faster than the truth on social media.

So, although the decades of investment in scientific research, infrastructure and innovation have undoubtedly paid off for humanity – the knowledge war threatens the very foundation of how science impacts the world. How? The COVID-19 vaccines are a good example. They were a scientific marvel. But because of the volume and speed of anti-vaccine misinformation via the Internet – many people will not take the vaccines – distrusting the science. Some polls suggest up to half of the population won't take the COVID-19 vaccines. The impact of science and innovation can be effectively neutralised because of misinformation. That is why I believe that misinformation is the single greatest challenge hindering human progress. But the knowledge wars are not a new occurrence for me, since I've experienced this intimately throughout my own 25 year research career as a climate scientist.

Starting out in my own scientific career, I assumed many things. I thought facts and evidence would change the world. The power of new data-driven insights around a beautifully constructed research project would be enough to change people's view of the world. I also thought a published paper in *Nature* or *Science* was the ultimate form of impact – it was vindication you were a top scientist in your field and the world would listen. In constructing my world view around publications and peers, I thought science communications to the media and general public was a complete waste of time.

Now I know that great science doesn't have a voice. It doesn't jump out from academic journals into the wider world. It must be communicated and promoted. It needs a narrative that grips people. But there was no training, incentives or guidance on the art of communicating science. That is still the case today. I was committed to this cause so, at the height of my scientific career – I resigned my well-funded tenured position to work on projects that help scientists share their knowledge so more people can learn from them.

Most of my peers think I'm crazy for giving up my academic career, but for me, science is as only as impactful as how well it is communicated and used by society. That wasn't the case generations ago. But with click-bait Internet and social media dominating how the world gets its information – science has to evolve in how it engages the world. We will lose the knowledge wars if we as scientists don't invest in communicating and engaging beyond our peers.

The old stuffy ways of communicating in jargon amongst ourselves don't work anymore. If we let old-school science communication flourish (jargon and university press releases), we are deluded in thinking we are doing a good job in translating our knowledge to the world. We are not.

The new science is not just doing innovative research – but allocating a good percentage of our time to training and communicating that knowledge beyond our peers. Whether to the media, school children, businesses or politicians – the new age of science is communication. If we don't, then the world will get its knowledge via click-bait sensational media, social feeds, Google searches and bloggers. That is what the world descended to in 2020. But that can change. The first step is for EVERY scientist to build the skill set to talk to anyone about their field. *Scientifically Speaking* is a brilliant resource to begin.

I believe all PhD students should be taught communication skills formally. While I would like all professors to as well – many are stuck in the old ways of communicating – happy and well-funded enough not to change – I understand. But at the minimum, the future generation should be formally taught as part of their studies in the art of communicating.

One thing I like to ask fellow scientists is if they have an elevator pitch summarising what they are working on in 30 seconds. If someone said "What are you working on?" would they be able to clearly understand it? And why it's important to them or society? I challenge you to do this. Then read *Scientifically Speaking* – which will help equip you with that basic skill. It sounds easy, but it's actually very hard given we tend to work in super-niche areas in research. You have to start by getting a wider perspective about your work. We tend to think everyone has a basic understanding of our field – but they don't!

I remember talking with Nobel Laureate immunologist Peter Doherty who was a mentor for me in my early career. Although I was in climate science and he was in immunology, he asked me to share a paper of mine with him during one of our meetups – as he had a curious mind and wanted to learn about climate. I shared with him one of my *Science* articles, thinking that would be a good introduction. He bluntly told me "I tried to read it, but your science is alien to me"! The language was so jargon-filled, it was useless to him. I was shocked at how bad my traditional science communications was if someone that brilliant couldn't understand a sentence of my own science – even if it was published in one of the most prestigious journals in the world.

He asked me to share a plain language summary for our next meeting. As scientists, we are hideously unequipped to communicate well beyond our fields but I strived to change how I communicated after that meeting.

As 2020 has demonstrated, it's insufficient for scientists to just stay in our labs and do nothing – from a societal perspective. Yet from a career perspective, sharing knowledge with a few peers is also history. Universities are encouraging and incentivising science communication and engaging with the outside world. In fact in my field, the most sought after scientists aren't necessarily the ones that do the most novel or innovative research – they are the ones who know how to communicate both amongst their peers, but most importantly, to those beyond.

Whether it's scientists in different fields or the media – or even social media – effective communication of science is not a choice – it's a prerequisite for all scientists. Whether to accelerate your own career or to make maximum impact on the world, realising the importance of developing and using effective communication cannot be understated. But how do we communicate to those outside of our fields? What is the art of communicating complex science to the broader world?

I've known Jo Filshie Browning for decades and she has taught me the skills to be a better communicator. When I'm interviewed on the BBC, writing opinion articles for popular magazines or sharing knowledge through social media, Jo has helped. Jo is passionate about science and equipping scientists to impact the world and it's a pleasure to read such a deep and insightful thesis of how we as scientists can better communicate.

Ben McNeil *is a long-time climate scientist, author and founder of Metafact.io.*

Speaking about science in the digital age

Now, more than ever, the scientific and medical community is under the microscope and in front of the media. Despite making incredible breakthroughs, the benefits of your work continue to be hotly debated. There's a movement of anti-experts, anti-vaxxers, climate change deniers and conspiracy theorists; fake news floods the mainstream media and flourishes on social feeds. This has grown worse during the COVID-19 pandemic, with fictions and falsehoods spreading as virulently as the virus itself.

The level of disinformation has grown because so much scientific discourse has now shifted online. When anyone with an opinion can log on and say what they want – whether or not it has any factual basis – it creates a lot of noise. It's as if there's a war being waged for people's attention, with scientific experts increasingly on the disadvantaged side.

What can you do to cut through the confusion and make yourself heard? First you have to acknowledge that to responsibly communicate science you have to understand and adapt to this new digital environment. It's no longer enough to write papers or even speak at conferences. You have to master the art of putting yourself across succinctly and engagingly online, which now means speaking on video in all its forms. The reality is that today it's not your 100-page research report or the thesis you spent three years writing that will make waves, but the 20-word sound bite, the media interview, or the online snippet of you discussing it.

Today, speaking is a non-negotiable element of communicating your science. And it's not just for the benefit of your subject, it will help you as well. When you apply for a research grant or a job, your name will be searched for online and the decision-maker will want to see your social profiles and judge how you come across on video. The next time a conference organiser is looking for a knowledgeable and impactful speaker, you want them to see your work on the internet and reach out to you.

Some experts have learned this already. They're invited to talk at congresses, are quoted in the media and find it relatively easy to gain support for their valuable views. They're respected, promoted and admired. So what's their secret?

After a professional lifetime of working with scientists and physicians around the world, I can tell you that to be truly successful in researching science and communicating it publicly, you have to prioritise becoming as good at speaking about what you do as you are at doing it. It's as simple – and as challenging – as that. If you don't have a clear, audience-focused message to impart at every level, whether it be on or offline, you're missing out. The experts I've described understand that science deserves to be communicated with clarity and vision so people will take notice of it, and they've learned to do it well.

Now you know this secret, there's no reason why you can't achieve the same level of success, but it will take a bit of time and effort to train yourself in the techniques. The truth is that you weren't taught how to speak persuasively and engagingly about your expertise at university or medical school. As a result, you've probably experienced the frustration of giving talks that didn't deliver the results you wanted, been misquoted by the media in a way that distorted your message, or tried to be invited onto a panel discussion only to be pipped to the post by someone with less expertise than you. Just as likely, you might never have sought the spotlight for fear of forgetting what you were going to say or coming across in the wrong way – which is a huge shame because your knowledge deserves an appreciative audience that will act on what you say.

This is where this book comes in. It covers the principles and practicalities of speaking about complex scientific information in the most compelling way, whether it be on screen, online or on stage. You'll learn how to deliver the right message, at the right time, to the right people and through the right channels, so you can not only serve science well but also your own career.

By the end, you'll understand how to make a genuine difference in the world through speaking about your subject. You'll also be able to influence public debate, gain more citations and enhance your professional profile; this will lead people to want to collaborate with you, which in turn will create more speaking opportunities in its own right. Most of all, your science will reach the audiences that need to hear about it.

I've seen this transformation happen with countless scientists and doctors during my 25-year career in healthcare communications, both as a practitioner and as an academic. My time as a journalist and media spokesperson in the healthcare sector taught me what the media wants to know, and my work as a media trainer has given me valuable experience in turning even the most reticent of professionals into confident speakers and communicators. What's more, I was a lecturer on the first undergraduate science communications course in the world and I have a diploma in scientific communications. Today I work with both senior and up-and-coming physicians and scientists as a coach and trainer, helping them to discover their inner communicator so they can influence others through their speaking. I'm also an international speaker myself.

In Part 1 of this book, you'll learn why speaking is so important and how it actually works (as opposed to how you may *think* it does). In Part 2, you'll master the process of creating and delivering a compelling talk that communicates your subject clearly and persuasively, along with how to handle a Q&A session with aplomb. And in Part 3, you'll discover how to speak engagingly in many different settings, while at the same time minimising your nerves and maximising your energy.

You may be reading this because you have a specific event coming up and want to jump to the relevant part, and that's fine. But I'd encourage you to read through the rest as well because you'll gain better results if you do. Becoming an engaging speaker means gaining an understanding of the fundamentals as well as delivering in a specific situation. My goal for you is that you'll use everything you learn in these pages to enable you to become the most compelling communicator you can be, giving talks and presentations that are as powerful as the science behind them. Good science should never be badly communicated, so let's make a start.

Part 1

Why you need to speak, and how it works

Chapter 1
The speaking advantage

Precision of communication is important, more important than ever, in our era of hair trigger balances, when a false or misunderstood word may create as much disaster as a sudden, thoughtless act.

James Thurber, American cartoonist and author

You're reading this book because you're curious about becoming a better speaker. But first it's worth asking why that's important. What's in it for you? When I talk to scientists and physicians from around the world, I'm given many reasons for why they want to speak with influence, but in the end they come down to two main ones:

1. To enable your research to make a bigger impact on the world.
2. To further your career.

These are worthwhile aims, and of course you may be motivated by both of them. You might also be asking yourself why you can't carry on as you are now, writing about your expertise and research projects and publishing them in scientific journals. Of course, specialist publications are still hugely important, as are scientific congresses and other science-led events. But they won't help you to win the war of attention that I mentioned earlier. Much has changed in the realm of communications in recent years, and the way you put across your messages needs to evolve to adapt to the current environment. Digital platforms have an impact on how you gain a profile at a global level, and audiences expect a different kind of communication than they once did. This means that what got you to where you are now won't necessarily get you where you want to be in the future; this new paradigm represents an opportunity, and to take full advantage of it you need to learn to speak with influence through a variety of different media.

What has changed

It was 3.00 pm on 21 October 2018 at a congress of the European Society of Medical Oncologists (ESMO), when a woman in a pale-blue dress walked onto the stage in front of a packed hall of 6000 people. Her name was Dr Kathleen Moore, and she was about to present the data behind a major breakthrough in ovarian cancer treatment: the Phase 3 results of the SOLO-1 study, looking at the developmental drug olaparib in women with the BRCA mutation of advanced ovarian cancer. As Dr Moore explained, in normal circumstances nine out of 10 women with the BRCA mutation die within four years, but this new treatment had been proven to achieve a 70% reduction in the risk of progression or death with olaparib versus placebo. For these patients, what we were essentially seeing was close to a cure. The high point of Dr Moore's talk was when she flicked a graph onto the screen behind her that illustrated the dramatic separation in the curves between the progression-free survival rates of the placebo group and the treatment group. At this moment there was an audible gasp from the audience, with tears appearing in some people's eyes as the full impact of the research came home to them.

The announcement was felt across the world, but it wasn't just because of the congress. In fact, Dr Moore's presentation was only part of the data communication. The work had actually begun the previous morning, when I'd chaired a press conference at which the key scientists involved had presented the research and then been interviewed by journalists. The resulting material was packaged up so that just as Dr Moore was walking onto the stage it was ready to be published in the medical media. Her presentation coincided with the publication of the study in the *New England Journal of Medicine*, and alongside this pre-prepared tweets were released by those involved in the study and others who were excited by the results. Images and video footage were posted online. It was a highly coordinated communications cascade, synchronised to hit simultaneously to create maximum reach. The result was a PR coup: articles appeared all over the world from the *Financial Times* to *Medscape*, various other media highlighted the 'breakthrough medicine',

Twitter went crazy, and Dr Moore and her colleagues from the trial had their science in the spotlight. I felt honoured to be a tiny part of this process.

However, leaving aside my own memories of the experience, what I want you to take on board is this: although many people saw the live presentation and some may have read the news articles, most found out about the research through social media, news websites and other digital channels. It's an illustration of how science communication has been going through structural changes that are driven by the increased prevalence of the spoken word in the digital sphere. A talk given at a conference is like planting a seed – it needs the soil and water of online media to bear fruit.

It wasn't always like this. Back in the nineteenth and twentieth centuries, the study of science was something that researchers carried out in small and exclusive groups. Writing and reading were the only channels for knowledge to reach other groups of scientists and the broader community. Academic papers were submitted and peer reviewed, and other experts might agree or disagree through writing letters to journals or conducting counter-studies; almost everything was conducted through the written word. Then, over time, international conferences were established so that scientists and doctors could communicate their work orally, which allowed a broader community who were fortunate enough to be able to travel to engage in in-person debate. As the internet revolution made its impact felt, blogs and online news articles were added into the mix, democratising scientific discovery and discussion across the world. And because it became increasingly obvious that, in order to be noticed, scientists had to surround their audiences with information, media ubiquity was born. Whether it be on Twitter, LinkedIn, news websites, or academic journals, their research had to be everywhere.

This worked well for a while, but the problem was that everyone started doing it. Today the challenge is that there are just too many words. Articles, papers, discussion forums, social media posts – who has time to read everything? Scientists are finding it increasingly difficult to achieve what PR people call 'cut through', which is when what you write is read by the people you want or need to see it. On top of this, you'll have noticed that we're all less willing to read now. We skim text as we scroll through our newsfeeds, taking in only the headline and summary; we 'read' books that are abbreviated to 15 minute cheat sheets or audio summaries; and we look at tweets rather than whole statements. So not only is there a huge amount of other written material out there to compete with yours, but the people you want to reach and influence are less inclined to read it in the first place.

What can you do about this? In a sea of written words, the *spoken* word now cuts through more than ever. And with the advent of the digital revolution, we can now speak with more people than ever before through online videos – far more than we can in face-to-face meetings. People like watching videos because if they can get what they need in a 30-second hit, that's far less demanding than reading an article. It's estimated that by 2022 online videos will make up more than 82% of all consumer internet traffic – that's 15 times higher than in 2017.[1]

The implication is that to win this war of attention you need to use your voice, and today that means accepting that video is king and that speaking on camera is necessary. Of course, printed media still has its part to play, but whereas once you would give a talk with a press release to go with it, now it must be accompanied by scheduled tweets and pre-recorded videos. And because conferences are filming, recording and live-streaming their talks – particularly post-COVID – it's safe to say that any scientific presentation you give will be disseminated much more widely than the walls of the congress room.

The bottom line

Speaking has become the key skill in scientific and medical communications, whether it be in person or on video. Think of the different ways you can communicate orally:

- presentations (both in-person and increasingly now online)
- congress TV
- press conferences
- panels (both in person and increasingly now online)
- poster sessions
- Twitter (this isn't oral, but the platform works much like a conversation pit and often includes video).

---

[1] www.forbes.com/sites/tjmccue/2020/02/05/looking-deep-into-the-state-of-online-video-for-2020

Does this scare you? If so, you're not alone! But I have good news, because speaking on digital media allows your work to reach a far wider audience than at an in-person conference. There are data showing that the greater the publicity around a publication, the larger the number of citations it will achieve, and the more significant its impact will be.[2] This is one way that speaking online gives you the opportunity to amplify your message.

If you're worried that you feel ill-equipped to put yourself across professionally when speaking, that's what this book is intended to help you with.

Furthering your career

If you're wanting to elevate your authority in your field, you live in exciting times. You have more opportunities than ever before to share your scientific research and message, and to bring people towards your way of thinking, whether it be through speaking at conferences, writing papers, recording and posting videos online or participating in online debates such as live panels or Twitter chats. These are all ways of winning the war of attention.

But here's the thing. Because times have changed, you must change with them. Your work does not end when, after labouring over a complex academic paper and submitting it to a journal, it's accepted. That's just the start of the process. If you want your work to be seen by as many colleagues as possible, if you want the best jobs, the best collaborations and the best funding, and if you want to recruit people into your latest trial, it helps if you can reach people through speaking. This doesn't mean you have to become like a celebrity, performing for the masses at their whim, but you do have to put yourself in front of the people who matter to you in the way that *they* want you to do it. This is something that you can learn to be good at, and if you're willing to do it, you'll be taking a significant step from being 'just' a scientist to being a highly regarded, expert professional.

To do this successfully, you have to be able to communicate well, explaining not just what you've done in your work but, more importantly, what it means for your audience. You also need to remember that you have two sets of people you want to influence: those in front of you and those who'll find out about what you say second-hand through digital channels. Suppose

---
[2] https://link.springer.com/article/10.1007/s11192-020-03380-1.

you're an expert in the reproduction of swell sharks and have some exciting research that you want to share with your peers. It's down to you to find a way of reaching them and making it interesting to listen to – in other words, it's your job to go to them rather than expecting them to find you. And if your ambition is to be a leader, this is what you must do: leaders speak in public because it's expected of them.

Making a bigger impact on the world

One of the most rewarding aspects of being a scientist is the opportunity you have to contribute to knowledge development – ideally to make the world a better place. Journalists and policy-makers are increasingly seeking scientific opinion to shed light on the hot topics of the day, such as public health (the COVID-19 pandemic being an obvious example), medical treatments and decisions, vaccines and climate change. They might want you to provide a scientific context for a decision, to legitimise a claim that someone is making, to explain a complex concept, to provide balance or a counter-opinion, to describe the implications of a discovery, or even to inform government policy. What vaccinations are needed during childhood in a specific country? What's the recommendation for a healthy diet? How should climate considerations drive energy policy? How should a pandemic be managed? We know only too well that the results of these decisions can change lives.

The stakes are so high in these scenarios, and the science so important, that when communications are handled badly it can have a catastrophic effect. A clear example is the MMR scandal in the 2010s, which associated the vaccine with the onset of autism and could be said to have kick started the current backlash against vaccination. Although the argument pitted one man, Andrew Wakefield, against the whole of the rest of the medical establishment, he was paid a disproportionate amount of attention. Why? Because he was one clear voice with a simple message. The numerous doctors and scientists who challenged him presented many valid counter-arguments, but they had no unifying central figure to put across their points and this weakened their messages. The situation was fuelled by a media tradition of giving 'fair balance' to a disagreement, which involved allowing two counter opinions equal airtime even though the 50–50 split didn't represent the views of an established opinion versus the voice of a single outlier. This might not seem logical to you – surely one person's opinion would gain less traction than those of hundreds? But that's not how humans (and the media) work. We like to focus on a single message, and prefer the consistency provided by a central viewpoint. In the case of Andrew Wakefield, what the scientific community desperately needed was a simple, clear, consistent, and memorable message to override his – preferably delivered by one key figurehead.

You may not aspire to be the next Robert Winston, presenting TV programmes and advising governments, but you can certainly be a respected and sought-after representative in your field. If you're known as a skilful communicator, you might be asked to present to a select committee, or to talk to a lay audience about what you know. And now, more than ever, your ability to communicate clearly and persuasively is critical because science has become so complicated. It matters that people understand it. Based on what you say, they make decisions about who to vote for, whether to vaccinate their children and what products and services they buy. You have the chance to influence human behaviour at scale, which is not only an honour and a responsibility but yet another reason why science deserves to be communicated well. Science doesn't speak for itself – you have to speak for it. And for it to be listened to and understood, you need to be an effective speaker.

Avoiding the pitfalls of digital immortality

Hopefully it's becoming clear by now that developing a positive and consistent presence for yourself online is an essential element of your influence as a speaker in the twenty-first century. When someone watches your presentation at a conference, or stumbles across your video, blog or tweet, their first reaction is often to find out more about you. They want to see you as a whole person, so they visit your LinkedIn profile, scroll through your Twitter timeline and check out any YouTube videos they find. What would they discover about you? Would they see someone who's a clear expert in a particular field or a confusing mismatch of content that gives an unprofessional impression?

It's worth thinking about this because, while a well-curated digital presence has the potential to spread your influence further than you could ever do by writing articles or speaking at the occasional conference alone, it follows that if there's anything that's less than professional on your social platforms it can last forever. This is something I call the pitfall of 'digital immortality'. As social media allow information to be shared quickly and broadly, any control by the person who created that information is often lost.

A couple of years ago I was at a science and technology conference when I saw the negative consequences of this in action. At a plenary session, the editor of a scientific journal remarked that he was being encouraged by his publisher to blog, but that he wasn't sure what a blog was. I could see jaws drop around me as people turned to each other in disbelief. Many in the audience were tweeting, which meant that the session was simultaneously being conducted both in the auditorium and online. As I scanned the tweets it became obvious to me that the panel members on stage, who were focused on the discussion in hand, were excluded from what had by now become the main debate. The comments were even picked up and tweeted by a mainstream science journalist, and appeared live on her paper's website. This is how an ill-judged aside by a senior member of the science communications community, made to a few hundred people at a congress, found its way into the mainstream media.

However, before you run for the non-digital hills, remember that being invisible online is not realistically an option today. This was demonstrated clearly during the lockdowns imposed on all countries during the COVID-19 pandemic, when physical conferences were cancelled. Many scientists who'd previously been reluctant to engage in social media took to sharing data, videos and articles on social platforms and open access journals in order to reach their colleagues as quickly and efficiently as possible. This was hugely beneficial in moving the emerging science forward. It also gave these scientists the opportunity to have their voices heard and to stay in touch with their peers. Those who hid behind their screens and didn't participate were, by contrast, invisible. If you want to have a voice, you need to be out there.

Nothing you learned at university about communicating science is helpful in the digital world

The way you learned to communicate when you were in a university setting is the complete opposite of how you should be speaking about your results to the public today. As a student, you were taught to communicate around the idea of the scientific method: to focus on the process of research and to explain it in writing following the same approach. It went something like this: 'I had a hypothesis, so I structured my study in this way and did a, b and c to research it. The result was x and so my conclusion is y and z.' That was fine for your supervisor, but fast forward a few years to now. What if you need to persuade a group of people to give you funding? Or to recruit subjects to participate in your research? Or to get your discoveries into the media so you can build a name for yourself or even change the national debate? To do this, you have to turn yourself from an explainer into an influencer. You need to focus on your audience, you need to be brief, and above all you need to be engaging.

You only have to look at television news bulletins and newspapers to learn how it's done. When you watch and read carefully, you'll be amazed by how much can be communicated in a short burst of time – a news bulletin can contain up to five stories in a couple of minutes. The media channels achieve this by understanding their audiences, gaining their attention and packaging the information in a way that's perfect for them to absorb.

That's why I'm inviting you to learn a new language: that of the media. It's even more important now that the people who used to be the 'translators' between scientists and the media – expert science journalists – are thin on the ground. Because of the rise of online platforms, there are fewer of them to do the communication job for you, so you need to take that task on yourself. Being fluent in this language is how amazing experts such as physicist Brian Cox have gained recognition – they're superb at explaining things and making people care about them. Like him, your job is to ignite the flame of curiosity and enthusiasm in your audience so its members care about it as much as you do – even for that one moment.

If you're wondering how on earth you're going to learn how to do all this, be aware that it's not a passive process; rather, it's one you have control over, and the secret is preparation, preparation, preparation. The people who are good at speaking and presenting are rarely those with an innate gift for it; they just put the work in, accept that they'll feel a bit uncomfortable to start with, and practise over and over again for many years. When you watch a great speaker, you never see the hours they've put into becoming that good. Believe me, I've worked with leaders in some of the most prestigious universities and companies in the world to help them with their performances, and they've spent hours doing it with me.

This is a skill that can be learned. Speaking is often stressful, but the irony is that the more you rehearse, the more spontaneous you can be, because you know what to do if something goes wrong. You can predict what the obvious questions will be at the end, plan how to answer them and be 'present' with your audience because you're not worrying about what you're going to say next. This allows you to respond to people's moods, treating them as useful feedback rather than as a worrying distraction. Even if you're talking to a camera, the confidence that comes with knowing how to present your information in an engaging way means you'll come across as relaxed and welcoming, which is so important for putting your message across. If you've ever sat through a presentation in which the speaker stares at their notes, a nervous sweat filming their brow, you'll know what I mean.

So often, scientists and doctors tell me that they're fearful of speaking because they distrust 'the media' – they worry that their words will be distorted or that people won't be sympathetic to their point of view. And this does sometimes happen. People might criticise you for doing experiments that involve genetic modification of plants or organisms, or working in stem cell research. They may disagree ideologically with you about your stance on climate science or vaccination. When you're in a public space, you're open to criticism and it's genuinely daunting – I get it. But if you're honest with yourself you'll realise that you might be worrying about an imagined worst-case scenario: in reality, it's unlikely that more than a handful of people will be hostile, if that. While you can't protect yourself from everything, you can go a long way towards receiving a positive response by making sure that what you say is clear, precise and unlikely to be misunderstood.

When you find your voice and feel comfortable with speaking, you'll realise how much of the process you have control over. There's also a virtuous circle involved: the more you speak, the better you become, the more you're asked to speak, and the greater the impact you'll achieve. And remember, if you don't speak up there'll be someone else, somewhere, ready to do it instead of you.

The main points

o Today, it's not enough to publish papers or speak at the occasional conference – you need to find ways of disseminating your work engagingly online.
o Learning to speak effectively can help you to further your career through making you more visible and influential.
o Speaking in public allows you to make a valuable impact on public life.
o Having a positive online presence helps you to avoid the pitfalls of digital immortality.
o You may not have learned how to speak effectively at university, but you can learn now.

Chapter 2
Setting the scene for speaking

Begin with the end in mind.
Stephen Covey, author of The 7 Habits of Highly Effective People

Speaking in public is enormously effective but it is also a difficult act to pull off if you don't know everything that goes into it. When we communicate, we're seeking to make a difference in some way: we want to move people, inform them, change their behaviour – or all three. It might seem like a transactional process with information passing from one person to another, but it's actually far more complex than that, with an infinite number of ways to get it right or wrong. In this chapter, we look at what communication really is, why it's essential to know what you want to get out of each and every event and the crucial differences between speaking to other scientists and to mainstream, non-scientific audiences.

It's not so simple

A standard communications model has been doing the rounds for many years. It's based on the idea that we transmit information from ourselves to other people via a 'medium', much like the way mobile phones send data to one another via a network. Difficulties can arise when the medium experiences interference, such as noise or static, but apart from that it's a pretty straightforward process. Figure 1 shows how it looks.

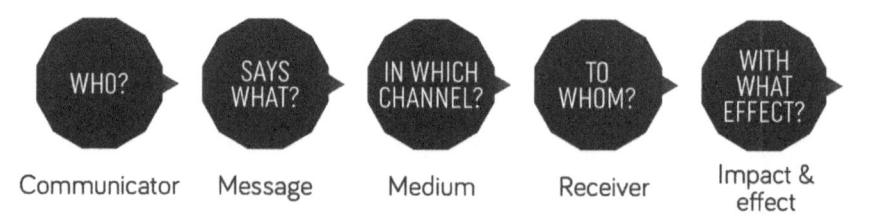

WHO?	SAYS WHAT?	IN WHICH CHANNEL?	TO WHOM?	WITH WHAT EFFECT?
Communicator	Message	Medium	Receiver	Impact & effect

Figure 1: The transmission model for communication

It appears beautifully simple, doesn't it? Some people even call it the 'communications equation', implying that everything balances out neatly at the end. However, the process of communicating is not simple, easy or balanced – in fact, it's highly complex and contains multiple variables, which can derail you without you even realising it.

How often do your listeners give you their full attention? Do you always communicate in a distraction-free environment? Can your audience always understand what you mean by the words you use? Does your tone of voice ever alter your meaning? Is your body language telling a different story from what you're saying? If you're thinking, 'That depends on the circumstances', you'd be right, because human beings are a capricious and illogical lot who interpret information based on what they're feeling in that moment. If only it were as easy as transplanting information into the brains of the people with whom you're speaking. The reality – and the challenge – is that it's more a case of you working out how best to convince them that what you have to say is worth listening to.

Where the communications model is helpful is in breaking down the process so we can analyse each part of it and see how it works (or rather, often doesn't), and where the pitfalls are. Let's walk through it.

Who? The communicator

This is you. You're a person who has lots of knowledge and an equal amount of assumptions, biases and preferences. You also have your own communication style, and vocal and physical traits. All this creates the potential to communicate correct information in a way that could be misunderstood.

Says what? The message

This is the information you're imparting. What you decide to say will be based on the kind of person you are, the circumstances in which you're delivering your information, and who you think you're talking to. There can be a whole lot of dangerous assumptions in there as well, as we'll see later.

In which channel? The medium

This is the means through which you're imparting your information, such as a presentation at a congress, a media interview, a Zoom meeting or a tweet. As you can imagine, the medium also complicates the process because it's a mediating tool – hence the name. It's an interposition between you and other people, so it's a key place in which transmission 'interference' can happen.

Imagine you're presenting your study at a conference and are scheduled to speak half an hour after someone else's talk, which contains a throw-away statement that undermines some elements of your hypothesis. This will have an impact on how your audience views what you say. Suddenly your work isn't definitive anymore, so the questions will not only be about your view but also about why your approach is different from that of the other expert. This 'interference' by the other speaker will have an effect on the impact of your own presentation, depending on how you handle it.

To complicate matters still further, you can add Twitter, a phone camera and computer screen, a journalist, a headline writer, an editor or any media channel into the mix – and you're still trying to communicate to people at the other end of it. If you're giving an interview, for instance, you have to remember that you're not really talking to the interviewer: you're talking to the congress TV viewers or the people watching the streamed programme on their computers.

What's more, when you're on stage (whether in person or virtually), it's easy to forget that you're also a medium in your own right, and one who may be experiencing an adrenaline or cortisol rush of nerves. So it's not just 'you', it's 'nervous you'; you're wired up, have a dry mouth and are liable to talk too quickly. This is why you need to take as much care of yourself as you do your message when you speak.

To whom? The receiver

This is the person or people to whom you're speaking. Earlier, I mentioned the personal baggage you have as a communicator. It's important to remember that the receivers at the other end of your communication have their own baggage as well. They may have strong opinions about your topic and the nature of your work, or a belief system that's different from yours. It's even possible that they won't have any scientific understanding at all.

It's easy to assume that your audience shares your language, and this can create endless misunderstandings; in medicine in particular, jargon can have a distancing effect. Your aim as a speaker is to bring your audience close to you so they can see things from your standpoint, not to push them away so they feel confused and alienated.

To add still further to the complexity, you actually have two audiences when you speak: the people immediately in front of you and those who will watch the video or read the report or media coverage that's generated as a result. This means you need to make sure that your message works not for you, but for all of your audiences. Everything you say should rest on who those people are, not on who you are and what you know.

With what effect? The end result

This is the impact you have on the people to whom you've spoken – the thing that they will go away thinking, feeling and doing as a result of your communication. It's your end goal, or the reason why you're speaking in the first place, and whether or not you reach it depends on how well you've understood yourself, your message, your medium and your audience.

The feedback loop

Another reason why the communications equation isn't sufficient as a model is that it ignores the presence of a feedback loop. If, part-way through your presentation, you look up and see a sea of blank faces broken up by the odd yawn, you have the chance to change what you're doing. Or you might be talking to a journalist who asks you to clarify something. Even if you're speaking on a video conference, you can usually see messages in the chat box or gain a sense of people's feelings from the expressions on their faces, and you can go back and re-explain something if you need to. Obviously you hope to get your talk right first time, but receiving feedback allows you another opportunity to put your message across in a different way.

The equation doesn't look so simple now, does it? You can see how many moving parts are involved, some of which can shift at the very last moment or even while you're speaking. That's why it is your responsibility to think clearly and systematically ahead of time. I believe it's your job to communicate your science well, not your audience's to understand, and to do that you need to remove as many of the uncertainties and variables around the communication process as possible. You can't make your talk perfect because there will always be someone who doesn't agree with you or isn't interested in what you say – that's life. But you can certainly give yourself the best possible chance of delivering your message so it's appreciated and understood.

Setting your goals

I've always loved this scene from Lewis Carroll's *Alice in Wonderland*, in which Alice says to the Cheshire Cat:

> 'Would you tell me, please, which way I should go from here?'
> 'That depends a good deal on where you want to go,' said the Cat.
> 'I don't much care where,' said Alice.
> 'Then it doesn't matter which way you go,' said the Cat.

Given that the communication process can be as challenging to navigate as Alice's adventures down the rabbit hole, you can see how essential it is to know what you want to achieve before you begin. If you're not clear on this, how do you stand any chance of inspiring your audience to think, feel or do what you would like? Having a clear road map for where you want them to go is the first step towards sending them off on the right journey.

Setting your goals is the very first thing you should do when you're planning your talk. Because after you've finished speaking, something will happen – whether you like it or not. Your audience might do exactly what you want, they might ignore you or they might question you. (They might not even see you in the first place if you picked the wrong medium.) Whatever the result, it's because you've created an impact and you want to make sure it's the right one. As an example of what powerful results speaking can achieve, witness how Italian physicians early in the COVID-19 pandemic were able to rapidly warn colleagues elsewhere in the world, via video messaging, to move virus testing units outside the hospitals to stop bringing infections in. The transmission of understanding through good communication can save lives.

When you define your goals, begin by asking yourself what success looks like to you. Write it down in as few words as possible, and bear in mind that you'll probably have more than one goal. For example, you could want to:

o share new information on the pathology of a disease;
o share the protocol of a new study on which you're collaborating;
o show that you have found a new way to target a challenging medical condition;
o share the data from a completed study about a new treatment regimen, and influence clinical practice;
o communicate the possible threats and opportunities faced by your organisation;
o show your peers that you're knowledgeable in a certain area; or
o have your opinion heard, respected and considered.

Let's suppose you're giving a presentation on the research results for a new medicine. One goal could be to ensure that its main benefit is understood so doctors in your audience can prescribe it appropriately. Another might be to influence health practice in the relevant area so the medicine is incorporated into guidelines in the right way. Another could be to acknowledge your research colleagues, and yet another to shine as an expert and possibly be invited into professional collaborations in the future.

I always suggest having goals that encompass what you want for your audience *and* what you want for yourself; it's important for your motivation that there is something in it for you. It's also helpful to allow yourself to say, 'I want to come across well', because this gives you an incentive to prepare and rehearse. You can make your message right, but if your mindset isn't also right, you won't serve the message well. As we know by now, a great message deserves a great messenger. This doesn't happen by magic – it takes a commitment to your self-development as a speaker.

So far, we've looked at examples of top-level goals, but you also need some smaller, more specific ones. Suppose your main goal is to influence leading oncologists, via an online presentation, to change their practice in a certain way. For them to alter their behaviour, you'd first need them to feel excited by your data. This would mean presenting it in such a way that those watching you at home stayed online until the end, then emailed you, asked questions and became part of your network. In addition, some lower-level goals could be to attract collaborators or recruit 1000 patients to take part in a clinical study. If you know your specific objectives, you can ask for what you want and therefore stand a good chance of achieving them.

Your goals also need to be realistic and achievable, which means tailoring them to the circumstances in which you're speaking. You might be giving 10 interviews at a congress about the results from a study into the efficacy of an experimental drug you've participated in testing, but what's your goal for the first interview? And the second? And the third? One interview might be with OncoAlert or Medscape, in which you're talking to a specialist health journalist who's an expert in clinical trials. Another might be with a reporter from a news agency such as Reuters or Bloomberg; they're less interested in how you developed the study than they are in the financial implications for the company that produced the experimental compound. Yet another interview might be with a journalist from the BBC, who wants to know about the patient impact and how long it will take for the study to come out, because this is what their viewers care about. Each interview is different: the first is about clinical practice; the second is about sales and profitability; and the third is about the impact on patients. One topic, but three different journalists and three different agendas. You can see how easy it would be for your message to come across in a confusing and inappropriate way if you didn't remind yourself at the start of each interview about your end goal for each.

One aspect to be aware of when goal-setting is the importance of being specific and concrete. Avoid using 'I want to create awareness' as an aim because it doesn't achieve anything. We're aware of many things in our lives, but we take action on very few of them, and there are some about which we may even have a negative perception. The same goes for 'I want to inform people', which amounts to much the same thing. Apart from anything else, creating awareness isn't ambitious enough – you want to stimulate specific change, not understanding for its own sake.

How you say it versus how the media want it

In 2018, a major study of various anti-depressant drugs was carried out and published in *The Lancet*[1] under the headline: 'Comparative efficacy and acceptability of 21 antidepressant drugs for the acute treatment of adults with

---

[1] www.thelancet.com/journals/lancet/article/PIIS0140-6736(17)32802-7/fulltext

major depressive disorder: a systematic review and network meta-analysis'. So far, so good. But look at the headline in *The Guardian*[2] newspaper: 'The drugs do work: antidepressants are effective, study shows'.

Nothing could indicate more starkly the difference between how the scientific and mainstream media talk about science. *The Guardian* article didn't even talk about the study, but led with the reassuring conclusion that 'antidepressants are not snake oil or a conspiracy' and that more people should be offered them. It's also worth noting the contribution of two doctors who were interviewed for the article. One of them, study leader Andrea Cipriani, didn't focus on how the research was carried out but instead clearly articulated the goal of the communication: 'antidepressants are an effective tool for depression … [which is a] huge burden to society'. The other, Professor John Geddes – who was not an author of the study – provided additional context and perspective: 'Depression is the single largest contributor to global disability that we have – a massive challenge for humankind.' Both these comments show that the focus was on showing *why* this study mattered to the paper's audience. The journalist responsible spotted what was interesting to a general audience and geared the communication appropriately. This was why it made an impact, and I suspect a lot of people changed their view of antidepressants and sought help from their doctors as a result.

News reporting is based on the engaging art of storytelling (we don't use the phrase 'news *stories*' for nothing). Reporters are trained to focus on the 'who, what, when, where and why' of a situation when they craft a feature, because they recognise that to grab people's attention they need to highlight what has changed and why it matters. This is what they lead with, and if you pay attention next time you read a news article you'll realise that the first thing you see is the 'so what' of the piece – the nugget that tells you why what you're about to read is important to the audience reading it. After that comes the 'what happened' part of the story, and finally the background information for those who are interested in learning more. In fact, there's a historical reason for this. Newspapers have always gained most of their revenue from advertising, and back when they only existed in print form, it wasn't unusual for a last-minute advertisement to be booked just before the

---

[2] www.theguardian.com/science/2018/feb/21/the-drugs-do-work-antidepressants-are-effective-study-shows

publication went to press. This would entail removing a chunk of news text as quickly as possible to allow space for it, hence the desire for editors to be able to lop off the tail end of a story while keeping the main points – at the top – intact.

Contrast this with how science is communicated in a professional or academic publication, or orally at a conference. The background comes first with the hypothesis for the study, then the design is shared and gone through in detail, followed by a reporting of the results and data – the 'what'. Finally, the conclusion gives the result – it comes at the end rather than the beginning. You can see the difference between the two approaches in Figure 2.

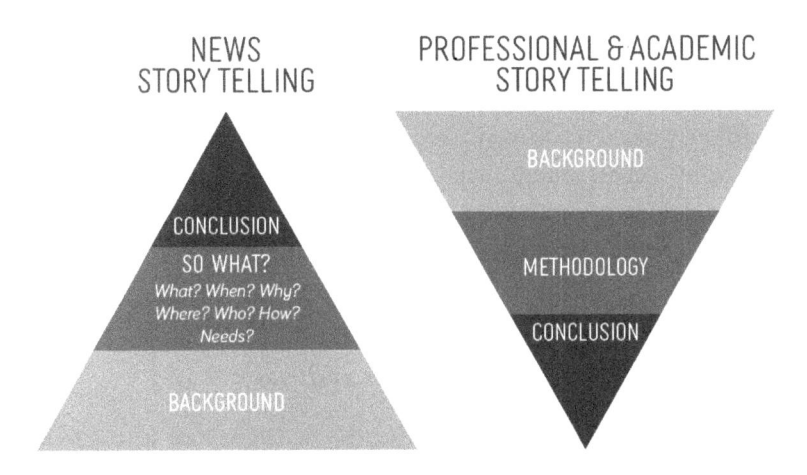

Figure 2: The media approach to storytelling vs the scientific and academic approach

Notice how instantly engaging and brief the left-hand version is. When you want to win the war of attention, you can take inspiration from this journalistic approach. To encourage people to listen to you, you need to home in on the 'so what' right from the beginning, and skim over the vast majority of the detail or return to it later.

Never forget that mainstream news has more in common with entertainment than information and education, and has a completely different structure and set of priorities from academic journalism. The former looks at the big picture, loves emotion and uses stories about individual people who their readers can relate to; the latter focuses on the detail, sticks only to facts and remains neutral.

Of course, if you were speaking in a scientific setting, you might want to follow some of the structure of the scientific method because you would be talking to your peers. But even then your presentation would be far more memorable if you were to announce your 'so what' at the beginning rather than leaving it until the end. The next time you have a discovery to announce, think about kicking it off with something like this: 'It's wonderful to have some good news here, because this is a special breakthrough in the treatment of lung disease.' It can take a bit of courage to break the mould, but consider how effective this would be not only on stage but also in the video footage taken at the conference; people would decide whether they were going to watch based on the first few seconds, and if it didn't grab their attention they'd click away.

Common pitfalls

You might be wondering how you're going to navigate this minefield without making some embarrassing mistakes. The good news is that this book is designed to help you avoid them, but the first step is to know what they are in the first place. With that in mind, here are the most common pitfalls to look out for.

Assuming that sharing information is sufficient

When you're communicating publicly about science, your job is to inspire rather than to simply inform, and this entails knowing who you're speaking to and putting yourself in their shoes. You need to meet them where they already are but not leave them there – take them on an inspiring journey so you can create change. Your talk shouldn't be a data dump, in which you tell people what you know and leave them to make up their own minds about why it matters. You want them to do, think and feel differently as a result of your talk – eat less red meat, take more vitamins, apply sunscreen. If they feel inspired, they'll act.

Knowing, and including, too much

Being choosy about your content for your talk or interview means jettisoning the vast majority of what you know and could include – never easy for someone who prides themselves on being an expert. I understand that this is challenging, especially when you're sharing something that's based on years of work. Wanting to showcase everything you've done is only to be expected. But this is where the difference between what you learned at university and what you're learning now comes in. In academia you gained marks for showing your working but in the public domain you're penalised for it. Non-scientific readers and viewers in the digital age only want a headline; the minute you start going into the detail or spouting too many statistics, you've lost them.

Not adapting your delivery to the digital age

There are many differences between the way you should present your information in writing and the way you should do it when speaking. Equally, many of the techniques and approaches for talking about your expertise on stage don't work when you're online or on video. There's an historical progression from writing to stage to screen which you may not have kept up with, and if that's the case you won't be making the impact you'd like.

Ignoring the fact that science isn't always precise

When you work in a scientific field, you're familiar with how knowledge constantly evolves. After all, if nothing ever changed doctors would still be advising their patients to smoke to clear their lungs. Yet, despite scientific understanding rarely being complete, decisions and recommendations have to be made. This can be hard for members of the public to understand, and it means that you have to be able to communicate uncertainties, risks and probabilities in an open and honest way but without confusing your central message. We'll cover this further in a future chapter, but for now be aware that this can be a pitfall.

Making it up as you go along

Hand on heart, as someone who was once a spokesperson myself and who trains people all the time to speak in public and give media interviews, it's possible to anticipate 95% of the questions you'll be asked in most situations *as long as you've done your research beforehand.* There's no reason for you to have unformed thoughts when it comes to dealing with challenges and queries about your presentations. When you think things through in advance and take the time to fully understand your audience, it pays dividends. This really is an area that's under your control.

The main points

- Communication is not a straightforward, transactional process – it's a sequence of causes and effects that can be distorted and interfered with every step of the way.
- Setting goals for your presentation, interview or panel discussion means you're more likely to get what you want out of it.
- The way you were taught to communicate at university and medical school is the complete opposite of how you need to communicate in the digital age.
- You can take inspiration from the mainstream media to learn how to communicate more effectively.
- Your responsibility is to inspire your audience to do something new, and to achieve that you need to understand who they are and how they want to be spoken to.

Part 2

The principles of persuasive speaking and presenting

Chapter 3
What to say

The biggest problem with communication is the illusion that it has taken place.
George Bernard Shaw

Good speaking doesn't happen by accident. It comes about because the speaker probably spent a lot of time preparing what they were going to say. It takes work – sometimes weeks' worth of work – to turn a complex subject into one that makes an impact at the level at which it needs to be understood.

When you spend your days in your own echo chamber, as we're all inclined to do, it can be hard to appreciate what it takes for people who aren't your peers to grasp the meaning of your work. The dangers of ignoring this are clear. Take the example of a drugs awareness campaign that was run a few years ago during Freshers' Week at a group of London universities. The aim was to educate new students about the effects of drugs and alcohol so they could make informed choices. To facilitate this, a survey was run that showed 90% of students had been offered one or more of these substances during that week. No surprise there, you might think, given that alcohol was included in the mix. The team working on this interpreted the result as a clear sign that there was a need for education about drugs; they believed students should have information to make choices, given the data. But the figure of 90% was irresistible to journalists, who took a different view: 'London students rush to drugs' screamed the headlines.

Imagine if something similar happened due to reporting on data from your research. I've seen results from a medical study showing a 60% response rate reported as 'Drugs chief says our drug doesn't work in 40% of patients'. The reality is that two people can look at the same piece of information and reach completely different conclusions, as shown in the cartoon in Figure 3.

Figure 3: Interpreting information can often depend on your perspective

The bottom line is that if you don't understand the perspective of the people with whom you're communicating, you can be misunderstood. To appreciate that perspective, you have to spend time thinking about who those people are and how they want to be spoken to.

In the previous chapter, you learned about setting your goals in order to get what you want out of your talk. In this chapter, you'll discover how to plan your message so it lands meaningfully with your specific audience and helps you to achieve those goals. Great ideas deserve great communication, so it's worth investing time and effort in tailoring your content appropriately.

Plan backwards

Whenever I start working with a new client who wants to deliver a brilliant presentation or give great interviews, I always ask them what they'd like to gain from our sessions. Invariably, they see this as a cue for them to explain what they want to say – in other words, to start with themselves. This is understandable because we do tend to focus on ourselves more than anyone else; we're always the most important person in our own minds. But when it comes to communicating, you can't afford only to think about what you want; in fact, you need to skip to the end of the process as your first step and start with your audience instead (see Figure 4).

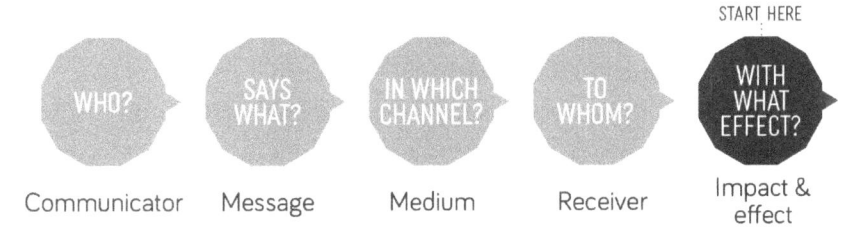

Figure 4: Start by asking 'With what effect?'

You've already set your goals, so that's the first stage complete. Then, if you move one step to the left, you'll see that the receiver is the next element to consider. We'll bypass the medium at this stage because we cover that in a later chapter, but after that we come to the message. Only when you know what that message is can you explore the role of the communicator – that's you (and how to develop yourself as a communicator forms the basis of the rest of the book).

This shows that you can't decide what you're going to say, or how you're going to say it, until you know and understand your audience. This is critical. Your messages and your audience are inextricably intertwined, and if they don't work in harmony you're in danger of confusing or alienating the very people you want to convince.

Your audience

Who are you speaking to?

The better you understand your audience, the more deeply your message will resonate with them. To begin this process, you first need to create a clear image of exactly who you're speaking to (see Figure 5).

Figure 5: Establish who you're speaking to

Usually this isn't too complicated – for instance, you might be presenting at a conference as part of a plenary session to a group of peers, or at a university to students. In many professional contexts, you're likely to be familiar with your audience because they're similar to you in many ways. But there will also be situations in which your audience is more complex. They may be unfamiliar to you, such as if you're presenting to a group of policy-makers for the first time. Or you may have multiple audiences to take into account – your talk might be filmed and live-streamed to a wider group of people with an interest in your topic, but who aren't in the room. In a media interview with a journalist your audience is actually their viewers or readers, not the journalist. Be clear about who you're speaking to, then make sure you do the work to understand them as deeply as possible.

No matter how well you think you know who you'll be speaking to, it's always a good discipline to 'check in' and make sure you're not making wrong assumptions. Have the people you're addressing at the forefront of your mind as you prepare your content. Ask yourself:

o   What do they already know about my topic?
o   What aspects of it will they be most interested in?
o   What's their level of understanding of it?
o   What vocabulary would they use to talk about it?

These are the main areas, but you can also use the following questions to broaden your understanding:

o   What are their pressing concerns?
o   Why do they need to know about this?
o   Why is this the right time for us to have this conversation?
o   Is there a related and topical event that they'll be thinking about right now?
o   Could they be hostile to what I have to say, or would they feel positive about it?
o   Why?
o   What are they expecting from me?
o   What do they want?
o   What don't they want?
o   Which words would have special meaning for them?
o   What topics do I want to emphasise or avoid?

o   What will persuade them to back my point of view?
o   What's the least they need to know?

You can't always assume you know or can imagine what (and how) other people think, so to do this properly may involve some online or in-person research. If you know the people you're speaking with, or others like them, you can conduct your own 'market research' by asking a few of them about their thinking on your topic. This will help you to understand their perspective in more depth.

It's also important to think about the environment in which you'll be speaking; it could be a symposium, a conference hall, a video livestream, an interview, a press briefing or a telephone interview. We'll explore these individual scenarios later in the book, but for now have a think about these top-line questions:

o   How long will I have?
o   What are the norms of the situation?
o   Will it be distraction-free, or noisy and crowded?

I hope you're starting to see how you can't possibly know how to deliver a persuasive message until you've put yourself in your audience's shoes. It's not enough to inform or educate them: your aim is to persuade them to take action on your goals. To do this, you must move and inspire them, or at least create a level of curiosity so they want to know more. I'm sure you instinctively understand this in any case. Suppose you were giving a presentation to some high school students with the aim of encouraging girls to go into STEM careers. You'd naturally tailor your content to a level they could understand. You wouldn't deliver a dry lecture at postgraduate student level and skip telling inspiring stories about successful women in science. If you want a specific outcome you need to begin with where your listeners are *right now*.

Your message

We're continuing to move to the left of the communications model (Figure 6), and now we come to your message (or messages, as we'll see in a moment). You're going to create one key message and four subsidiary ones, but for now let's just focus on the main one.

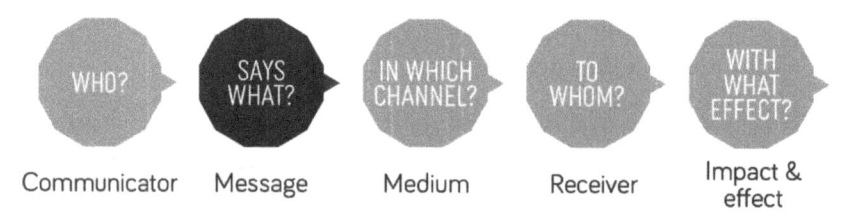

Figure 6: Next focus on your main message

Again, we're going to take inspiration from journalists and use the questions they ask themselves as a way of harvesting the basic information for your talk:

o   Who?
o   What?
o   When
o   Where?
o   Why?
o   And ... so what?

Write your answers longhand or in a mind map, keeping them as clear and concise as possible – they should be top-line rather than detailed. Remember that you're basing them on your goals and also your topic. This is just a brain dump to help you get the information out onto paper – think of it as a harvesting technique to create some 'headlines'. It's an excellent way of ordering your points, and it gives you a hierarchy of information – what's most interesting and important versus what's least significant.

Tailoring your message to your audience

In Australia, as part of a campaign to reduce the level of skin cancer, research was carried out with focus groups to understand what information would encourage people to use sunscreen. It seemed that the risk of skin cancer wasn't a key driver – in fact, people didn't want to think about 'dying' when they were on the beach. They were keen to look good and get a tan. But when probed more deeply, it seemed that vanity could be a useful motivator for messages supporting the use of sunscreen. The very same people who were turned off by the 'dying from the sun' message were also those who cherished their appearance and didn't want to look old before their time. So the advertising agency created a visual of sets of twins, some of whom had lived in Australia's sunny climate for years and others who'd lived in the United Kingdom. Those who'd lived in Australia looked decades older than the others, and this had a motivating effect on sunbathers. The goal was to persuade them to wear sunscreen, and the message helped to achieve it. Although this example is based on visual images rather than speaking, the point is that attaining a successful result depends on the communicator both knowing what they want to achieve and truly understanding their audience.

Do you want your listeners to appreciate what you're telling them, or do you want to tell them what they'll appreciate listening to? The first is 'you' focused and the second is 'audience' focused, just as it was in the Australian ad campaign. It's the second one that you should aspire to. It might take time to discover the special something that will move your audience, but don't skip this step. Find your core message and tailor it to them before going any further, because it gives shape and direction to everything that follows. Without it, you run the risk that you will ramble on, veering from one unrelated topic to another, and look up from your notes only to see people scrolling through their phones in search of something more interesting.

Getting your message across in a way that works for your audience can seem like a complex task, which is why I've come up with a three-step process for planning it out. It's pretty foolproof, so I highly recommend that you follow it.

1. Plan your journey

Looking at what you know about your audience, ask yourself three sequential questions:

1. What are they thinking about my subject right now, if anything?
2. What do I want them to think if I'm to achieve my goals?
3. What would take them from where they are now to where I want them to be?

Let's say your goal is to encourage as many people as possible to get their flu vaccinations. Right now it's not on their radar, or if it is it seems like an unpleasant inconvenience that's not worth the hassle. To achieve your goal, you want them to think, 'This is a priority for me and worth making time for – I'm definitely booking a shot tomorrow.' But what would take them from A to B?

This is where your audience research comes in. For most of your audience, it's probably not the thought of falling ill that's the motivator, but the desire to protect their family from infection and serious consequences, or to avoid taking time off work – both the result of being ill. More recently there may be the additional benefit that flu vaccination helps in the management of COVID-19, as those who've been vaccinated have certain strains of flu ruled out as a potential differential diagnosis. These are the things that matter to your audience and are the areas to focus on if you want to encourage them to have a shot, not the research on the vaccine or the success rate percentages. (Of course, if your audience was made up of scientists rather than the general public, your message might be different.)

In another scenario, you might want to encourage people to feel more curious about an area of your research, or to change their medical practice. Alternatively, you may be talking to people who are critics of your work, or who have strongly held preconceived ideas about what you're presenting. In this case, you'd need to fully understand their perspective and arguments so you're able to challenge their assumptions without alienating them. You adapt your message to suit who you're speaking to.

2. Play some games

Here's where you can have some fun – in fact it's compulsory. The purpose of games is to loosen up your thinking and encourage you to make new mental connections between audience and message. Try these options:

- o If you were writing a newspaper headline about your talk, what would it be?
- o If you were going to summarise your talk in one sentence, what would you say?
- o Could you do it in a tweet?
- o If you were going to search for your talk on Google, what would the question be?

You can see that the process of answering these questions not only means thinking about your message from different perspectives, but also cuts out extraneous detail. It's essential to do this to clarify what you want to say.

However, your answers still might not be completely right for your audience, so when you've finished ask yourself whether you should tweak them. How can your main message be more relatable? Do you need to change the vocabulary, or to sound less erudite? Do you need to keep it formal, or can you be less 'scientific'? Keep going until you're happy with it.

3. Build your temple

So far, you've focused on one core message, but clearly you'll need more than just one to convince people, however long or short your speaking opportunity. You must have a structure, and this is where it's helpful to think of your messages as making up the elements of a Greek temple (Figure 7).

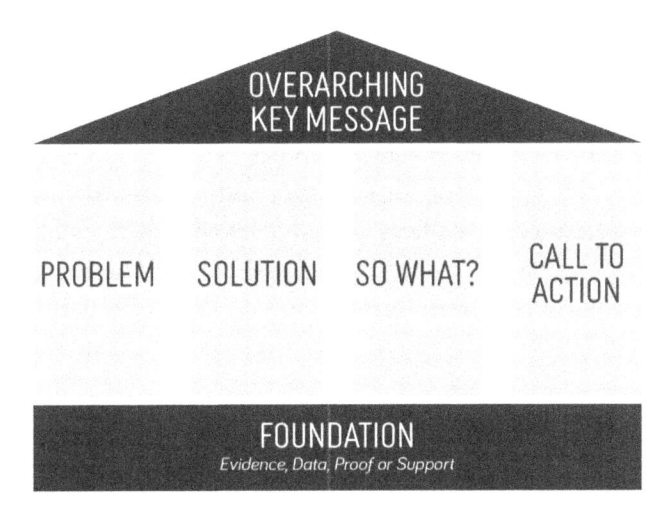

Figure 7: Your messages in the form of a Greek temple

The top, triangular part of the temple is your overarching message (the one we've worked on already), which dictates everything that comes beneath it. Underneath that you have four pillars containing your subsidiary messages, and under each of them you will find some data or evidence to back up your arguments.

o   *Pillar 1:* an explanation of the problem you're solving through your talk.
o   *Pillar 2:* a summary of the solution you've come up with. Note that for a non-scientific audience you've skipped the details of your work entirely here; they're less interested in the structure of your study than in the implications of what you've discovered.
o   *Pillar 3:* what the solution means for your audience (the 'So what?'). What can the science do for them?
o   *Pillar 4:* your call to action (what you want your audience to do as a result).

Here are a couple of examples so you can see what I mean.

Example 1: If I were speaking about this book

- *Overarching message:* 'Great science deserves great communication.'
- *Problem:* Speaking impactfully and accurately about science can be challenging – especially in the digital age – a situation made worse by the fact that effective science communication skills are not taught as part of science and medical degrees.
- *Solution:* You can learn to master the art of speaking well, equipping yourself with the skills and confidence to communicate effectively and authentically with all audiences.
- *So what?* When audiences are engaged, science can be advanced, systems can be enhanced and patient outcomes can be improved. Now more than ever, it's important for scientists and doctors to cut through the noise, communicate the facts and connect with their audiences.
- *Call to action:* If you're a scientist speaking in public on any platform, you would truly benefit from learning how to do it effectively if you want to enhance your career and advance the cause of science in the world.

Example 2: If you were speaking about quitting smoking

- *Overarching message:* 'Giving up smoking can be the single best thing you can do for your health, and to be successful you need as much medical support as possible.'
- *Problem:* Smokers are at far greater risk of diseases that affect the heart and blood vessels than non-smokers. Smoking causes strokes and coronary heart disease, which are among the leading causes of death around the world. To add to this, giving up smoking is hard because nicotine is so addictive.
- *Solution:* You need both willpower and medical support if you're to give up.
- *So what?* If you want long-term success as a smoker, you should seek help from a health professional and make a plan to quit.
- *Call to action:* Make this your year to give up smoking – sort out an appointment with a health professional today and make a plan to quit.

When you have your messages planned out, stand back and take a look. It's tempting to add large amounts of detail, but resist this: short and simple is best. Remember that your aim is to be interesting and persuasive, so in your hierarchy of information you should focus on the information that will be most motivating to people; this might mean leaving out elements that you find fascinating, but that would cause your audience to be bored or confused. Also keep your points concrete rather than theoretical, jargon-free, simple, clear and consistent. What's in it for your audience? Why would they care?

At the base of the temple are the foundations that support the pillars. These are the facts and figures that underpin your talk: the evidence from studies, clinical trials and epidemiological data. These foundations are critical parts of your messaging because without them your information is just assertion, so think about what evidence would make what you say most believable and authoritative. Having said that, the data are mainly there to support your argument rather than to take centre stage. Examples of how you could talk about the data are: 'We know this because the data showed …' Or 'What's exciting in the data is …' It's there as a backup rather than as the main point.

Of course, you would have a different temple for each audience, so if you're speaking on the same subject to divergent people, you'll need a temple for each.

Don't lose the emotion in the message

You shouldn't think of your messages as facts alone – there should also be emotion embedded within them. I often speak with people who are adamant that science is not the place for feelings. Of course, it's true that science tends to be fact based, and some audiences respond well to this, but I'm passionate in my belief that there's a definite need for emotion in the *communication* of science.

Many people are surprised and uncomfortable when I ask them what they want their audience to *feel* as a result of their talk. I think this is because they think of emotions as being dramatic – the type of intense feelings we see at the cinema or on stage, such as jealousy, rage or love. But the kind of emotions I'm talking about are more 'professional' than that, such as pride, suspense, anticipation or hope. In fact there's a huge range of human emotions that you can stimulate in your audience, and for inspiration you can use the Plutchik Wheel of Emotion.[1] Developed by Robert Plutchik, it analyses all the feelings we have and breaks them down into eight basic ones:

---

[1] www.6seconds.org/2020/08/11/plutchik-wheel-emotions

joy, trust, fear, surprise, sadness, anticipation, anger and disgust. The wheel also shows how many variations there are on these, and visualises them for you so you can use them as a way of imagining the effects that you might create.

I'll talk more about the importance of emotions in the next chapter, but for now just make sure you're taking them into consideration when formulating your messages.

The main points

- To speak persuasively, you must start with your audience and then tailor your messages to the people who are part of it.
- To do that, you need to understand and empathise with those you're speaking to.
- An effective talk requires a strong structure, which is what the Greek temple approach gives you.

Chapter 4
How to say it

I've learned that people will forget what you said, people will forget what you did, but people will never forget how you made them feel.

Maya Angelou

When you're speaking, your aim isn't only to put across your information so that it's understood, but also to enthuse people so they care as much about it as you do. This is a key way of winning the war of attention. How will you make your messages matter? How will you share your passion? To do this, you have to *connect* with your audience by packaging up your information in a form that's palatable to them so they can't help but remember it and take action. This is important because the stakes are high. The situation in which you're speaking could be one of many – a poster, presentation, panel discussion or interview – but whatever it is, you can assume that today there's a high chance you'll be recorded and the video shared online. So it's not just the people in the room watching you but a potential audience of thousands.

The truth is that being an inspiring speaker isn't a dark art that only a few can achieve. There are many techniques that you can use to bring your speaking to life, and in this chapter you'll learn how great speakers use rhetorical structure and storytelling to engage their audiences. You'll also discover techniques such as framing, analogies, sound bites and statistics to make your messages both understandable and memorable. If all this feels a little overwhelming, just pick one technique to use for your next talk and see how it goes. Then add another one, and another. Effective science communicators have a whole toolbox of approaches on which they can call at any time, but they don't use all of them on every occasion.

Before we begin, there's one thing I need to make clear: a high-stakes discussion or media interview is not the time to think spontaneously. You need to allow time to prepare thoroughly if you're going to make an effective impression. I often find that people put weeks into creating their slides, but barely any effort into planning the talking points around them. This is doing things the wrong way around, because what you *say* about your visual aids will stick in people's minds far longer than your graph on slide three. It's the way you tell it that counts.

Ancient wisdom

The ancients knew a thing or two about speaking, and rhetoric is no exception – in fact, the word itself comes from the Greek word for speaker. Hundreds of years ago, Aristotle and his contemporaries had almost the opposite problem that we face: reading and writing were elite skills and most communications were delivered in person. The goal for them was to speak so that people would remember what they said. They had no cameras or audio, so they relied on creating memories. This is how Aristotle developed rhetoric, which is the art of speaking effectively. Essentially, it's the oral communications 'operating system', and it consists of three pillars: *ethos*, *pathos* and *logos*. We can learn from rhetoric even now.

o *Ethos:* 'Believe me because of who I am.' When you cite an established expert or accepted behaviour such as scientific research principles, you're using *ethos*. It's about your reputation, expertise, capabilities and reliability. *Ethos* causes people to listen to you because of who you are and what you represent.
o *Pathos:* 'Believe me because I feel your pain.' When you arouse feelings in people, you're using *pathos*, and as we know strong emotions have the power to align people towards a single purpose. *Pathos* also creates empathy (the ability to imagine things from another person's perspective) and encourages people to take what you say to heart.
o *Logos:* 'Believe me because I have evidence.' People love simplicity and predictability. When you make an argument with a logical structure, it's impressive, and when you support your case with facts, published research and studies, you give further weight to it. *Logos* is an appeal to

reasonableness; if a lot of work has been done in an area, it's reasonable that your audience should take it seriously.

Fast forward from Aristotle's day to now, and you can see how successful communication still relies on all three of these elements working together. The relative importance of each depends on who you're talking to, your message and the situation you're in, but you still need all three. If you're talking to scientists, their primary interest is likely to be *logos*, but if they're to believe you, they'll also want to have evidence of your expertise (*ethos*) and will be persuaded by emotion (*pathos*). If you want donors or volunteers to support your research, you'll want to prioritise *pathos*, but you must still be a credible person (*ethos*) and support what you're saying with evidence (*logos*). You can use *ethos, pathos* and *logos* as a checklist to make sure you've covered all areas when you're planning a talk.

Emotion is like superglue

Although there's an important place for all three elements of rhetoric, I find that the one most overlooked in scientific speaking is *pathos*. For this reason, I'm going to expand on the importance of emotion so you can see what a transformative effect it has on your talk.

I included the Maya Angelou quote at the start of this chapter for good reason. When you're speaking, you're a human being talking to other human beings, not a data-point talking to other data-points. If you're to move your audience to think in a new way or to take an action they wouldn't have considered before, you need to make them *feel* something. What's more, emotions activate the limbic part of the brain, which is the primal area that creates memories. That's why talks that spark an emotional response are 'sticky', hanging around in people's heads for weeks, months or even years. Science shows us that emotion and memory go hand in hand.

As I've mentioned before, journalists are experts in capturing attention by creating human angles for stories. Several years ago, I worked on the publication of a study by the mental health charity SANE in Australia. It was an excellent example of emotional storytelling in science. The charity's aim was to shift government policy towards giving more support to patients with what was at that time an under-appreciated psychiatric problem: bipolar disorder. So it commissioned a study called Bipolar Disease Costs, which examined the ways in which poor funding of the disease was costing taxpayers more in the long run than subsidising it well. The conclusions within the study itself were based on the data, but the article headline ran thus: 'Bipolar Disorder Cost Report reveals the shocking economic and human cost of neglecting bipolar disorder.' The body of the piece quoted the figures, but its most compelling aspect was an interview with a man who had bipolar disease. He described it as 'like living in a hurricane belt'. Every time he had an episode, it was as if his whole life was destroyed, then he'd rebuild it, only for it to be devastated by another one sweeping through. It was heartbreaking – I still remember it today. The point I'd like you to take from this is that when you're giving a talk, you can help people to remember and appreciate your messages far more easily when there's emotion involved. No one remembers a statistic unless they care deeply about it.

As a scientist, it can be hard to move away from a reasoned, intellectual approach to speaking because it goes against the scientific method that you were originally taught. Scientists rightly tend to favour *logos*, and I'm not suggesting that you abandon logic altogether – especially if you're presenting to your peers. In those situations, you need it just as much as you ever did. However, you can still add an emotional perspective by starting and ending your talk with an explanation of why it matters so much to you to be there. And if you're being interviewed on congress TV or online, you still have the opportunity to tell the human story behind your research: 'We've been working on this development for five years, and to be able to share it today and see the reaction from everyone is amazing. We want to make sure the data are disseminated as widely as possible and can't wait to bring this new treatment to patients.'

It's as important for you to have a specific emotion that you want to generate as it is to have a set of facts to communicate. Remember the ovarian cancer presentation I described at the beginning of the book? The feeling that came from that was hope fulfilled, along with the confidence the study gave to doctors who wanted reassurance that they would be doing the right thing by prescribing the medicine when it became available.

Now we've established how important feelings and memories are for your audience, we'll look at various ways in which you can create the right ones (and avoid the wrong ones) when you're speaking. It all comes down to how you put your content across, and we'll start with techniques that involve how you structure your talk. These are:

- framing
- storytelling, and
- 'the rule of three'.

Next we'll look at how you can use words in different ways to create the impact you want, through:

- analogies, and
- sound bites and statistics.

And finally, we'll cover off the two areas you want to avoid if your talk is to achieve the effect you deserve, which are:

- jargon and 'science-speak', and
- the passive voice.

Framing

This is how you help your audience to think about your topic in the way you want, and to do it well you need to know your audience. When speaking with non-expert audiences in particular, this is an essential part of communicating science responsibly. You're telling people about both the 'what' and the 'so what?' – in other words, what the audience needs to know about the data and how they should interpret the findings. Think of it as a 'before and after' for your data, creating a context for the information. What's obvious to you may not be obvious to others, so it's your job to frame it for them so they know what to do with the information you've shared. Information without context is an invitation for people to create their own interpretations, so by framing what you say, you're using your expertise to help them to navigate it.

Consider how you want your listeners to think, feel or act. Will they find the new data about a targeted therapy exciting because these results promise hope for a new treatment? Are the statistics about COVID-19 worrying, therefore requiring people to increase their level of personal vigilance? I remember one transplant surgeon telling me how important framing was when he was first talking about transplanting pig valves into human hearts. The most common reaction from the public was 'yuck', so the surgeon told me that every time he spoke about it publicly, he used the phrase, 'This is good news, because it will help save lives.'

You can use framing phrases to lead your audience in the right direction, both before and after your information delivery, such as:

o   I'm hoping that what I'm about to share will change your views on …
o   We were concerned about why so many patients were suffering from this disease, so we …
o   The backdrop to this is that …
o   What's reassuring about this information is …
o   This is a significant development because …
o   This is good news because …

You may feel uncomfortable about influencing people through framing because you consider it your job as a scientist to be objective, enabling people to make up their own minds. However, even when you're talking to other experts, you still have a responsibility to help them understand your message by putting it into a context they can appreciate. This is doubly so when talking to a lay audience. To non-experts, framing is an important way of explaining things that could easily be misinterpreted or misunderstood.

How does this work in reality? Imagine you're presenting research to a group of doctors that shows widespread screening for colorectal cancer reduces mortality. Your aim is to encourage countries to adopt early testing if they don't already do so, so here's how you could frame your data:

> *Before:* 'As colorectal cancer is one of the most common and lethal cancers in Western countries, we wanted to understand the impact of screening programmes by undertaking a meta-analysis of many large-scale studies from around the world. The results were conclusive, clearly showing that screening led to a decrease in colorectal cancer mortality rates and making the case for widespread use of screening approaches.'

This sets up your audience to listen to your information, after which you would give the body of your talk. Then at the end:

> *After:* 'So we've demonstrated that if you want to reduce bowel cancer death rates, screening programmes are hugely valuable. They detect cancer earlier, allowing for better treatment which leads to better outcomes.'

This gives the 'so what?' – the prompt that encourages people to take the action you desire.

It's so simple, isn't it? It provides the context for a talk and whets the appetite of your audience. Yet so often it isn't done, which represents a missed opportunity for both speaker and audience.

Storytelling

Picture the scene. Rolf Heuer, CERN's Director General, takes to the lectern in front of a packed auditorium. 'It's a special day,' he says, and goes on to deliver a long and highly technical presentation illustrated with slides showing equations and various data. As the talk reaches its conclusion, the presenter intones: 'If we combine the ZZ and the gamma gamma … They line up extremely well in the region of 125 GV and they combine to give us a combined significance of five standard deviations.' The audience erupts into applause. More graphs follow from another presenter, accompanied by yet more clapping. The camera zooms in on Peter Higgs, the man responsible for the hypothesis of what's now known as the Higgs boson, who wipes a tear from his eye. A microphone is handed to him and in between deep breaths he quietly says, 'I would like to add my congratulations to everybody involved in this tremendous achievement. For me, it's really an incredible thing that has happened in my lifetime.' A murmur of appreciation sweeps the audience as the modesty of his statement sinks in, and yet more applause fills the room.[1]

When I watched that video, I was wiping away a tear as well. And it wasn't because I knew anything about particle physics; it was because I could appreciate the weight of the moment and what it meant to those involved. It even prompted me to want to learn more about it. Why? Because the video told an emotional story. The main characters were the scientists, their quest was to discover the Higgs boson, and the ending was the fulfilment of a life's work. It wasn't the physics that was at the centre of it, but the people.

Storytelling is one of the most effective tools you can use when you're wanting your message to stick – there really is nothing like it! The reasons for this are well researched.

Stories are engaging

Given that we know the importance of emotion when speaking, we have also to acknowledge the importance of stories in generating that emotion.

---
[1] https://youtu.be/0CugLD9HF94

This is because they alter our brain chemistry. When we listen to a story with fear or suspense at its heart, we release cortisol (the 'stress hormone'), which gains our attention. When we want to know what happens next, we release dopamine (the 'reward neurotransmitter'), which prompts us to learn and remember. And when we hear a story in which we empathise with the characters, we release oxytocin (the 'bonding hormone'), which brings us close to the storyteller and encourages us to trust them.

These chemicals create a special phenomenon that occurs when we listen to a story, called 'narrative transportation'. This is when we imagine ourselves as being part of the tale, identifying with the situation on which it's based and sharing the feelings of the main character who drives it. It's why everyone enjoys listening to stories – after all, we pay good money to watch them at the cinema. People are more likely to hang around to hear what you have to say if you tell them a story than if you just throw facts at them.

Stories are memorable

Because stories are organised in a consistent way with a beginning, middle and end, they provide a helpful way of structuring material. This also makes them memorable – up to 22 times more so than facts alone.[2] If you want your scientific message to stick, a story is a far better way of helping people to remember it than simply imparting the data. Think of stories as little anchors, embedding your pieces of information into people's brains.

Stories help with understanding

Stories help us to get our heads around new concepts because they're how we've been interpreting the world since we were born (indeed, since the dawn of time). They're the key way in which we create sense and meaning out of seemingly chaotic situations. We're wired for stories – we listen out for them, and we're constantly trying to form our own stories as well.

You can see that stories speak to how we like to be communicated with – they work with human nature, rather than against it. I'm sure you know this from your everyday life. If you come home from work and your partner were to ask how your day went, you wouldn't list the number of coffees you drank or the emails you sent, but you might recount the tale of how your grant

---

[2] https://womensleadership.stanford.edu/stories

application had been rejected yet again, along with who'd been involved and why. Or you might explain how a colleague had been promoted, and how happy you felt for them. In other words, you'd tell a story.

Stories are about people, not statistics, and this makes them the perfect vehicle for emotion-based information-sharing. They don't have to be on an epic scale, and they can be professional as well as personal. Instead of simply showing a slide full of figures, you can tell a story around it as well: 'When we looked at the data, I'll be honest, we were stumped. But when we delved more deeply, something surprising emerged …'

So how do you tell a story? There are many types of plots and narratives, but there are three that I use the most. They're tried and tested, and are deeply satisfying for audiences to hear.

The hero's journey (*The Hobbit, Star Wars, Harry Potter*)

> A scientist has a theory, and after overcoming countless challenges over many years, eventually discovers something amazing. This is a way of describing your research so that it means something to people. It brings out the personal story of the science, presenting the analysis as a quest with hurdles, false starts and disappointments; finally all are overcome and success is achieved.

Craig Venter gives a TED Talk in which he describes sequencing the genome.[3] He starts by saying:

> We're here to today to announce the first synthetic cell, a cell made by starting with the digital code in the computer, building the chromosome from four bottles of chemicals, assembling that chromosome in yeast, transplanting it into a recipient bacterial cell, and transforming that cell into a new bacterial species … This is a project that had its inception 15 years ago when our team … was involved in sequencing the first two genomes in history. Could we understand the basis of cellular life at the genetic level? It's been a 15-year quest just to get to the starting point now to be able to answer those questions.

He frames the process of discovery as a quest, then proceeds to give an overview of the journey of discovery that was to be fraught with failures

---

[3] www.ted.com/talks/craig_venter_watch_me_unveil_synthetic_life

and setbacks. Then finally, success. 'Early one morning, at 6.00 am, we got a text from Dan saying that now the first blue colonies existed!' The success is presented in the context of the people undertaking the process of discovery, and their persistence and determination in the face of adversity. Do you see how he uses both storytelling and framing here? The super complex science of cells and genetics becomes a quest of discovery, and I'm hooked.

This structure also often works well for science that has a negative or equivocal result, by putting the lack of a positive trial outcome into the wider context of the journey of discovery. 'While we'd hoped for a positive outcome from the study, and are disappointed that this isn't something that's proven successful for patients, we have pushed the research further along. The study did show the validity of these biomarkers in a large population. It also gives us a clear path forward when it comes to designing trials in this area.' In effect, this reframes a negative event into a broader and more positive context.

Now and in the future (*Tomorrow's World, An Inconvenient Truth*)

> A scientist develops something (possibly a theory) that changes how the future will be. This is a helpful way to talk about the 'so what?' of a discovery – what it delivers to people and its implications. You start by describing a problem and the issues it creates, then show what the world would look like with the problem solved. In between you can talk about how to get from one to the other, which naturally incorporates your work or expertise.

Think of precision medicine as an example: 'Now we understand the genetic markers, just imagine how cancer treatment will be in the future. You could have a blood test which would tell you what kind of cancer you were at a high risk of getting, and what you could do to prevent it.'

This type of structure can also work well as a warning about what will happen if nothing is done. The book *The Future We Choose: Surviving the Climate Crisis*, by the scientists involved in the Paris Climate Accord, uses this structure with powerful effect.[4] The authors describe in detail two contrasting examples of what the world could be like in 2050, one optimistic (if action is taken now) and the other catastrophic (if aggressive climate policies are not adopted).

---

[4] Christiana Figueres and Tom Rivett-Carnac, *The Future We Choose: Surviving the Climate Crisis*, 2020.

Circular (*Forrest Gump, Talking to Strangers, The Secret Life of Walter Mitty*)

> A scientist describes the beginning of their journey and what happened along the way, and at the end brings the listener back to the beginning. This can be useful as a way of explaining something complex, or talking about a new discovery. You could start with a simple situation, such as a decision, a diagnosis or a scientific phenomenon, explain it in detail, then revisit it at the end with the more experienced perspective that you have by that time.

For example, a doctor talks about how they first became fascinated by heart surgery at medical school, then goes on to describe the challenges of becoming an expert in the cardiac field. At the end they return to their student days to compare what they knew then to now.

> I'm amazed to think back to that student of 30 years ago, watching open heart surgery for the first time. And to think about what's been discovered and achieved in a relatively short professional career since then.

The key elements of a story

Storytelling is one of those techniques that, once you get the hang of it, becomes easier and more enjoyable as you go along. But to start with it's helpful to understand that every story must have five key elements:

1. a message or moral
2. characters
3. conflict
4. a series of events (a plot), and
5. concrete, sensory details.

Use these points as a checklist when you're creating your story to make sure you've covered all bases. In particular, don't forget the sensory details because those are what fire people's emotions. In my workshops with clients, I often use this to highlight the power of stimulating the senses:

> Imagine I'm holding, right here in my hand, a bright yellow lemon. I'm going to put it down, and I'm going to cut it in half, and then I'm going to cut another wedge out of it. You can see that it has this thick skin – it's a beautiful, ripe, Italian lemon. Then I'm going to ask you to take a piece and bite into it.

What's happening to your mouth? If you're anything like my clients, you're salivating. This is how powerful your imagination is in changing your physical and mental state, and it is the reason why painting a picture for your audience is so effective in engaging their attention.

The rule of three

If plunging into storytelling seems a bit of a tall order for now, another helpful way of structuring your talk is to use the 'rule of three'. Just as we humans are hardwired for stories, we're also programmed to think in threes. Think of the Holy Trinity (father, son and holy spirit); *veni, vidi, vici*; the good, the bad and the ugly. There's something satisfying about this format, and because it's been around for so long it helps to anchor what you're saying into people's memories. As the Romans used to say, *omne trium perfectum* (everything that is three is perfect).

So how do you use the rule of three for your speaking? There are (appropriately) three main ways, and they're not mutually exclusive:

1. You can divide your talk into three parts with a beginning, middle and end. This is a simple and obvious structure, and it is used when presenters start with an introduction to what they're going to say, say it and then sum up what they've said.
2. Have three elements that you explore. For instance, if I were giving a talk about how to prepare a presentation, I might cover messaging, body language and mindset. You might talk about a scientific problem, how you addressed it and what this means for the future.
3. Because things that come in threes are so memorable, you can use the rule to create a sound bite:
    - Never in the history of human conflict was so much owed by so many to so few. (Winston Churchill during the Battle of Britain)
    - Homes have been lost; jobs shed; businesses shuttered. (Barack Obama's inaugural speech)
    - Stay at home. Protect the NHS. Save lives. (UK Government at the start of the first UK lockdown in 2020 during the COVID-19 pandemic).

This technique helps you to structure your content so it's easy to understand and, more importantly, to remember. It also has the advantage of preventing you from cramming too much information into your talk because you have to limit it to three areas.

As a summary, we've explored three ways of structuring your content: framing, storytelling and the rule of three. Now we'll take a look at two interesting ways in which you can use language strategically to reach your speaking goals.

Analogies

So often in science, you're trying to communicate abstract concepts or highly complex material, which can be difficult for your audience to grasp. Analogies are the perfect way to make such concepts more accessible because people don't have to know the science to understand them, they just need to get the analogy instead. If done well, they can be a powerful shortcut to help create understanding. And what's more, good analogies are highly persuasive because through their very nature they tap into people's emotions.[5]

Let's look at a couple of examples. Several years ago, I worked with a doctor who was preparing to speak to the media about a new medicine that used phytodynamic therapy to target and kill skin cancer cells. At that point, it was a new therapy and I was keen to understand how a cream was able to be so specific. How could it tell the difference between healthy and cancerous cells? The doctor explained it to me by way of an analogy:

> It's like weedkiller. Just as weedkiller is formulated to take advantage of the fact that weeds grow more quickly than grass, so this cream targets the cancer cells, which grow more quickly than the surrounding healthy ones. The cancer cells 'grab' the drug from the cream and absorb it more quickly than the others. When you shine a light on the skin you can see it's the cancer cells that are full of the drug, and they're the ones that are killed.

Here's another one. In 2004, astrophysicist Neil deGrasse Tyson appeared on the NBC *Today Show* to talk about the $3 billion Cassini orbiter probe that was sent to study Saturn and its moons. The host challenged him on the enormous cost – was it worth it, he asked? Tyson's view was that it was, but he had to find a way to put it persuasively. First he explained that the expenditure was spread out over a dozen years, and that this amounted to less than Americans spent annually on lip balm. While this didn't convince

---

[5] https://hbr.org/2005/04/how-strategists-really-think-tapping-the-power-of-analogy

his interviewer (who pointed out that it was taxpayers who were funding the probe), it did elicit cheers from the viewers standing outside in the plaza and was the subject of social media coverage because it was so unusual. Most importantly, it put the cost into the kind of context that Tyson wanted – that of a cheap, everyday purchase. But Tyson still had to convince the host and his viewers that the money was going to a worthwhile cause, so the next analogy he used was to liken space exploration to gaining a greater understanding of what he called 'our neighbourhood'. 'We're not an island here,' he said. 'We're part of an interacting system of comets and asteroids, and what we learn about the rest of the planets tells us all kinds of things about what our past and what our future might be.' By using the word 'neighbourhood', Tyson brought the solar system into viewers' backyards, helping them to appreciate its relevance to their lives in a vivid and memorable way.

It would be hard to find a more masterful use of analogies than this. But did Tyson think of them on the spot? Of course not. Later on he explained, 'I didn't pull that out of thin air ... I went in with 10 different calculations. I know the audience and the interviewers. I pulled the lip balm out of my utility belt. You can't just say exploration is good ... Compare it to something else. That's the tactic. I have people to this day telling me about the lip balm moment.'[6]

Analogies are powerful, but they must be used with care and a degree of precision because an inappropriate one can be damaging. For example, many have criticised the fact that when COVID-19 first started spreading, it was often compared to the flu. Unlike flu, COVID-19 is a respiratory illness that also has implications for the heart, lungs and blood, so comparing it to flu perhaps meant that the general public (and some politicians) didn't take it as seriously as they could have done. In contrast, I was delighted when I saw one doctor say on social media, 'Why would you want anything that turns your blood into lumps of jelly?' What a graphic and easy-to-understand analogy that communicated the impact of a serious case of COVID-19!

So how do you find and create your own powerful analogies? It can be challenging, especially if you're not used to thinking creatively and laterally, but the more you do it the easier it becomes. Ask yourself whether there's

---

[6] www.forbes.com/sites/carminegallo/2020/02/12/neil-degrasse-tyson-teaches-three-persuasive-communication-strategies-in-his-new-masterclass

anything that's similar to what you want to explain, but in the everyday world. You could find inspiration in the areas of transport, household objects, computers, the natural world, relationships, business, politics – anything. Breaking it down also helps. Take the end result and think of the steps you could take towards achieving it, thereby creating an analogy for one step rather than the whole.

Craig Venter described creating an artificial cell by emptying it and putting in new genetic material that sparks life as 'booting it up with new software'. And I've heard an epilepsy treatment described thus: 'When someone has a seizure it creates a scar in the brain, as if someone has walked through a cornfield and left a path. The drug allows the corn to grow back so the path is forgotten, and in the same way the brain forgets to have the seizure.' You can see how relating scientific concepts to concrete actions that everyone can relate to makes them both memorable and easy to understand.

Sound bites and statistics

Another way of using the right words to create the effect you want is through the use of sound bites and carefully selected statistics. The beauty of these is that they're intensely repeatable, feeding into the 'Did you know?' mindset; your sound bite could end up being tweeted, talked about at a dinner party or even repeated back to you by a colleague.

Again, sound bites don't tend to come naturally – they need preparation. When Oscar Wilde declared, 'I can resist everything but temptation', I'm sure he didn't come up with it on the spot. Achieving simplicity is often the hardest thing to do in any area of life. When I worked in the field of smoking cessation and wanted to get across the addictive nature of nicotine, I found a few interesting facts and quotes, including Mark Twain's assertion that giving up smoking is easy because 'I've done it hundreds of times'. I also learned that only three to five of 100 smokers who give up cold turkey will still be non-smokers a year later. These facts were helpful, but finally I settled on the fact that research showed 'nicotine is more addictive than heroin or cocaine'. That sound bite really brought home how hard the habit was to kick.

As for statistics, as you will remember from the previous chapter, facts and statistics are the foundations of your temple of communication. The key to making them effective is to present them in a way that means something to your audience – it's like a combination of framing and analogies (think of Neil deGrasse Tyson and the lip balm). If you're talking about a disease, how many people have the illness globally or in your country? If you're presenting research, what will the impact be? How many people can benefit? Then think about the long-term picture. Will there be a cost to introduce this intervention? What's your view? Is it money well spent? At all times, try to compare it with something that's relevant to the world of your audience, making the numbers human and easy to relate to. So, 'This programme will save one million people from blindness each year' is followed by, 'which means that seven million people won't lose the breadwinner in their family to a preventable disease.'

We've looked at two ways you can engage and motivate your audience through language: analogies, and sound bites and statistics. To round off, we'll cover two further ways of speaking that will help you to win over people's hearts and minds.

Talk in 'human-speak'

I was in my office one day when the phone rang. It was a woman from a young, dynamic company based in my home town, and she explained that she wanted my help with creating a launch plan for an incubator. She was contacting me, she said, because of my public relations experience in the medical field.

The more we talked, the more enthusiastic I became: one of my own children had spent time in an incubator after he was born, and although I didn't mention this, I listened and asked questions. While I was doing so, I reflected on the fact that one of my team's fathers was a paediatrician specialising in pre-term babies – there were all sorts of ways we could bring his expertise into the project. This was going to be good, and I could already imagine how the plan might roll out.

At this point, the woman mentioned something to do with speaking opportunities at finance expos. This puzzled me, and while we were on the phone I Googled 'finance expo'. When I saw the results, I stopped speaking for a moment and stared at the screen. She wasn't talking about baby incubators, but about business start-up incubators, or investment companies that funded new enterprises – a completely different concept. Luckily I hadn't given away my misunderstanding and was able to extricate myself with grace, but it was a close shave!

I'm sure we've all had similar experiences to this, when we've been foxed by language that can mean different things to different people, or that's incomprehensible in the first place. Unfortunately, it's often scientists who are most guilty of causing the problem. There are few professions that love technical jargon more than the medical and scientific community, and there seems to be a 'members club' in which only those who are in the know can understand what anyone's saying. More often than not, it's the unintended result of spending one's working life describing things in specific terms to other people in the field, but it's easy to forget that not everyone can relate to it. Can you point to your sternocleidomastoid or the tip of your olecranon? Unless you're in the medical profession, probably not. And why would you?

We can assume that it shows intelligence to talk in obscure terms, but actually it's more of a talent to be clear. So what can you do to achieve clarity?

Avoid jargon as much as possible

Make sure your vocabulary is pitched at the right level for your audience. If you're presenting peer to peer, you can use the technical terms your informed audience will understand, but if you're speaking to scientists from a different background, or to people who don't have your expertise, modify them accordingly. Don't assume that other people have your knowledge.

Having said that, bear in mind that depending on your goals, you might want to educate your audience so they can understand your language. It could even be that they want to learn it so they feel better informed. The best way to achieve this is to use the jargon and then explain its meaning in a non-patronising way – for instance, 'the technical term is a myocardial infarction but for most of us it's known as a heart attack, and it happens when blood flow to the heart decreases or stops, causing damage to the heart muscle'.

A lovely example of this is in Shakespeare's *Macbeth*. Macbeth has just murdered Duncan and feels overwhelmed by guilt. Looking at the hand that did the deed, he says:

> Will all great Neptune's ocean wash this blood
> Clean from my hand? No, this my hand will rather
> The multitudinous seas incarnadine,
> Making the green one red.

Shakespeare is subtly explaining to his audience that 'incarnadine' means 'turn something crimson or red', but it's not patronising – it's educative. In the same way, you can preserve the integrity of your science by using technical language and also avoid the distancing effect it creates by 'translating' for your audience. You want to find the overlap in a Venn diagram of vocabulary that works for you, your audience and your profession.

Avoid acronyms and abbreviations

Acronyms and abbreviations can be just as distancing as technical jargon, because they mean something to an exclusive circle but not to everyone else. (And just so I'm practising what I preach, I'll explain that an acronym is an abbreviation formed from the initial letters of words and then pronounced as a word, such as NASA – the National Aeronautics and Space Administration.) What's even worse about acronyms is that they can mean different things to different people. Consider these examples:

o ATS: American Thoracic Society or Applicant Tracking System
o ESMO: European Society for Medical Oncology or European Science Mission Operations
o ASH: Action on Smoking and Health or American Society of Hypertension or American Society of Haematology
o ESC: Eurovision Song Contest or Embryonic Stem Cell or European Society of Cardiology

You can see the potential for misunderstandings if you don't explain your acronyms and abbreviations. Of course, if they're well known – such as VW, BBC, EU or IBM – they're fine as they are.

Speak directly to your audience

In the scientific method, you're taught to use the passive voice: '200 research subjects were given the maximum dose' or 'the trial was ended after a certain number of events had been reached'. The active voice versions of these would be: 'Our team gave 200 research subjects the maximum dose' or 'We ended the trial after we'd reached a specific number of events'. Can you see how the former creates a distancing effect and the latter is more engaging and involving? This is because the third person, which is used in the passive voice, gives the impression of a 'God-like' persona. This may work in clinical papers, but real life is about people not things, so I encourage you to use the active voice as much as possible when you're speaking about your work.

In addition, the active voice helps you to be clear because it shows who's doing what. The passive voice doesn't identify the person or people performing an action, and obscures who's responsible for it. This is why politicians say 'mistakes were made' rather than 'I'm sorry, I made a mistake'.

Having said that, there are occasions on which you could use the passive voice to good effect. The first is when you're speaking to a highly scientific group of people who would expect it (although even then there's a place for the active voice at the beginning and end of your talk). The second is when you want to distance yourself from what you're saying. For instance, if you're presenting the results of a drug that has a rare but serious side-effect, you might want to depersonalise the situation and talk in the passive voice: 'It was noted in the trial that this side effect was reported.'

What I'd like you to take on board is that choosing the passive versus the active voice should be a deliberate choice based on your goal. You shouldn't be slipping into the passive voice throughout your talk simply because it's what you're most familiar with, so keep practising using 'I', 'we', and 'you' until it becomes second nature.

Putting everything into practice

You've learned a lot of techniques in this chapter, and have much to think about. Which should you choose to focus on first, and which would work best for your audience, so you stand a good chance of winning the war of attention? Here are some questions you could ask to help you decide:

o   Does my message need framing in a particular way?
o   How does my view differ from the accepted one?
o   What common ground do I have with my audience, if any?
o   How will I inspire my listeners?
o   What language should I use and avoid?
o   How am I going to stimulate my audience's feelings and win them over?
o   Am I avoiding the passive voice in favour of a more direct style?

It's also worth writing out your talk before you practise speaking it, because it's a great way to clarify your content. If you end up with more to say than you have time for, don't worry, because it allows you to be flexible to suit your audience on the day. You'll be following in the tradition of the world's great performers; the comedian Robin Williams used to say that he had 10 hours of material for every two-hour show. Most importantly, by knowing everything you plan to say ahead of time, you'll be prepared for any eventuality and this will help you to feel more relaxed when it comes to the event itself.

The main points

- To inspire your audience to do what you want, they must feel something. No talk is complete without emotion.
- The laws of rhetoric encompass all the elements of a credible and inspiring talk.
- Framing is a technique you can use to help people to interpret your facts.
- Storytelling is a tried and trusted way of winning people's hearts, minds and attention.
- The rule of three is a helpful structuring device that promotes clarity and understanding.
- Analogies, sound bites and statistics give you ways to make your messages entertaining and memorable.
- Generally, you want to avoid using jargon, acronyms and the passive voice, as they create a distancing effect.

Chapter 5
Mastering the Q&A

Who has questions for my answers?

Henry Kissinger

So far we've covered the proactive elements of speaking: the content you want to put across and how you intend to deliver it. But how do you deal with the reactive part? This could be the Q&A at the end of a presentation, an interview with a journalist or a conversation as part of a panel discussion.

However, is it as reactive as you think? My take on this is different from that of many people: I believe that the way you answer the questions should be prepared for *just as thoroughly* as your presentation. In fact, I'd go so far as to say that in many cases the Q&A is *the* most important part of the presentation, and there are a number of reasons for this.

o *It's an opportunity.* Handled well, you can use your answers as a way to continue the story you gave in your talk and to re-emphasise your messages in a different context. Think of it as a second bite of the cherry.
o *It can derail you if you're not careful.* If you're asked a question that's related (but not central) to your theme, it's easy to wander off topic. Before you know it, you're miles away from what you wanted to focus on. Suppose you've given a presentation about a clinical trial protocol, and during the Q&A you're unexpectedly asked about a different but related trial that has just been completed. You answer the question about the other trial,

then one question leads to another, and somehow you end up talking about a subject that you haven't worked on directly and don't know much about. This can not only make you look ill-informed, but also remove the opportunity to emphasise the points you originally wanted to make. And this is before we even get into the dangers of answering challenging questions the wrong way.

o *It's at the end.* Being the last thing that people hear, your answers are what they will most remember. If you don't do a good job in the Q&A, you could be left with the unpleasant feeling that you've wasted an opportunity or even diminished the impact of your talk.

o *It's highly shareable.* You're always on the record. Even when the Q&A is off stage, you can be recorded answering questions. (Journalists don't take notes anymore, they simply hold a recorder up to you.) Some members of the audience at a talk or panel discussion may even be filming you, so you only have one chance to get your answers right. Often the most exciting and tweetable moments of a presentation are in the questions and discussion because they're the spontaneous and interesting snippets. That's good news if you give a compelling answer and bad news if you put your foot in it; you might end up winning the war of the wrong kind of attention.

Given that there are so many good reasons for preparing for the Q&A, I'm constantly amazed at how little attention people pay to it. I think scientists fall into two camps with this. Some assume that it's the easy bit, when the pressure's off and they can simply react to what they're asked. Others are so terrified about the prospect that they'd rather not think about it at all. It's as if they see the Q&A as an expanse of shark-infested sea, with great whites ready to tear them to pieces if they dip a toe in the water. They feel it's much easier to focus on the element they *can* control, which is the main body of their talk.

You can see how these two viewpoints, while at first seeming contradictory, have one thing in common: there's an underlying assumption that the questioners hold all the cards. However, this is not the case. As the speaker *you're* in control, and you're about to learn how to make a Q&A, interview or panel discussion work for you. I should add that, for speaking opportunities, this requires a certain mindset shift away from thinking of it as an add-on that you'll deal with on the day to an integral and equally important element of your talk that you can and should prepare for. It's up to you to make it a success.

The questions you might be asked

The Q&A or an interview is an opportunity for your audience to engage with you. If you have a poster, you need to have a short pitch statement to give to whoever's stopped by. If you're giving a presentation, there will be questions at the end (even if it's online). If you're on a panel, your performance will revolve around the audience questions and the discussions these generate with your fellow members. And if you're giving a media interview, it's all about the questions.

From my experience of being both a journalist and a spokesperson, and of preparing clients for numerous Q&As, the questions you'll be asked fall into four areas (see Figure 8). You won't necessarily be exposed to all four types, but you need to be prepared just in case.

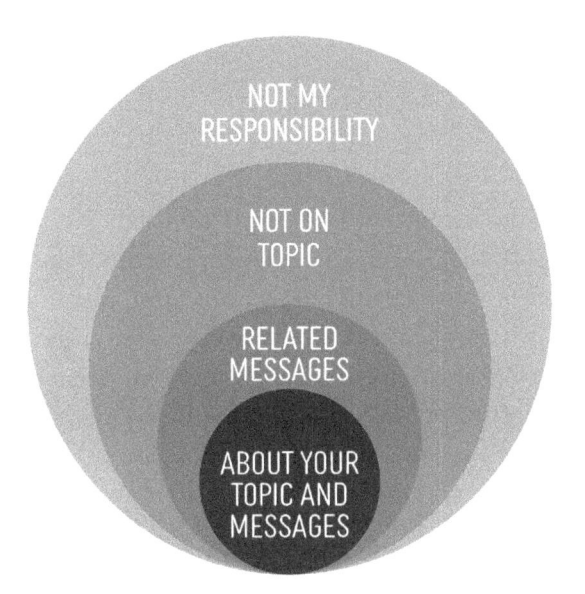

Figure 8: The Q&A – your topics

Questions about your key messages

These are what I call 'gift questions' – the ones that lead you straight to talking about the messages you've embedded in your talk. Examples are:

o   Tell me about your study.
o   How important is this research finding?
o   What do the results mean for the audience watching or listening?

If you're asked this kind of question (and it will usually just be one) it will probably be the first in an interview or panel discussion, as the host or journalist wants to warm you up before probing further. And what a question! It's as if your interviewer has handed you a gift with a bow on top, inviting you to talk about the very subject you're there to discuss.

However, just because it's easy doesn't necessarily mean it's a breeze to answer. You have to give a neat and snappy reply that tells a potted version of your story right from the beginning, a bit like an elevator pitch. Think of it as having three parts:

1. what you did
2. what it showed, and
3. what it means.

Then reverse the order like this:

1. what it means
2. what it showed, and
3. what you did.

You want to focus mainly on numbers 1 and 2 in the reversed list, the 'meaning' and 'showing', which are what people are most interested in. Only emphasise number 3, the 'doing', for a medical or scientific audience, and even then only if you think they'll be interested. Be aware that because it's such an obvious and easy question, people will be unimpressed if you don't do an eloquent job of answering it. I recommend practising saying it out loud again and again so it's embedded in your memory.

The following three question types are more challenging to answer, so I'll outline them here and then spend the rest of the chapter explaining how to answer them.

Questions about areas related to your key messages

These are on topic, but stretch it further. They could involve exploring the detail, delving into the 'so what?' of your work or trying to establish the broader context. Examples are:

o What are the side-effects or downsides of this treatment approach?
o This sounds like an expensive treatment. Do you think it will be affordable?

- How does your study compare with similar ones?
- What can people at home do about this?

These can be more challenging, as they require you to extend your thinking beyond the core messages you've developed yet still give an appropriate answer while maintaining a good impression. To do a good job of this, you need to think about and possibly research your answers ahead of time. For instance, you should know what the side-effects and downsides of the treatment are and how many people have had them, or when the treatment is likely to be available. You have to have your answers ready if you're to avoid looking ill-informed or being led off track.

Questions that are off topic

Here we're getting closer to the shark-infested waters I mentioned earlier. These questions are vaguely related to the subject you want to talk about, but are designed to lead you far away from it. Examples are:

- How does this relate to [the issue of the day]?
- Given that many medicines are 'financially toxic' do you think this drug is too expensive?
- This only provides an incremental benefit. Is there ever likely to be a cure?

The most common version of this type of question is one that asks you to comment on a topical issue that's only tangentially related to your expertise. For instance, if a lung disease expert were on a radio show at any time from March 2020 onwards, they'd have been asked about the handling of the COVID-19 pandemic. They might not have been there to talk about it, but they'd still have to deal with it in such a way that it brought the discussion back to their expertise. If you're ever in a similar situation, you'll find that the further away you travel from what you know, the less comfortable you'll feel, and the less likely you are to come across as authoritative.

A friend of mine was a journalist in Scotland during the Scottish referendum, and at the end of every single interview, whatever the topic was, she'd ask, 'So, how are you voting?' She told me that at least three times the answer was so interesting that she ran with it as the main headline. Her interviewees assumed it was just a throwaway question at the end, but for some of them it overrode everything they'd said beforehand. Beware of this.

Questions that aren't your responsibility

These are questions that it is legitimately not your job to answer, and to handle them confidently you have to have pre-set boundaries about what you will and won't say. Examples are:

o   What do you think about [a different research team's] progress in this area?
o   What would you say to the Prime Minister about the way they've handled healthcare funding?
o   Why is [someone else's paper on a different subject] so inconclusive?

At first these might seem relatively easy to answer, in that you can say, 'Not my problem.' Except you can't, because you'll come across as evasive. Nor can you take the risk of speculating or commenting on something that could derail your story and become the controversial take-out message of an interview or Q&A, overshadowing all the other messages that you're seeking to communicate. You want to give a professional and well-informed impression, so you have to anticipate these questions and think of an elegant way of responding to them.

You can see that most of these question types pose challenges in which you have to find a way of answering them professionally but without causing problems for yourself. I wouldn't blame you for feeling nervous about answering them, but shortly I'll walk you through how to handle each type of eventuality. For now, just appreciate how essential preparation is for dealing professionally with the Q&A.

Grasping the opportunity

First, I'd like you to focus on what a golden opportunity the Q&A is to drive home your messages again and again. It's not the time to be passive and answer them meticulously as if you're trying to get an 'A' grade in an exam, but rather to step into your authority as a speaker and steer the conversation in the direction of your choice. People need to hear information repeatedly and in different contexts before it sinks in, and the Q&A is your chance to help them understand it. It doesn't mean becoming a slippery politician, but it does mean learning a couple of techniques that will help you out.

Bridging

This is a trick that accomplished media interviewees use to stay on track. It's when you give an answer to a question and then build a bridge to the answer you *really* want to give – the one that delivers one of your messages. For instance, if you were asked the gift question, 'Tell me, how large was your study?' you wouldn't just reply, '140 sites across 28 countries' even if it sounds impressive. You'd also say, 'This analysis looked at more than 120,000 people, and the great thing is that because we have such a large amount of data we can feel completely confident that our conclusions are robust and tell us [insert key message here].' Even an easy and relatively mundane question is an opportunity for you to give your 'So what?'.

But what if the question is more challenging? You can still use bridging, but in such a way that it steers you to back to what you want to talk about. For instance, if you're asked why your trial has taken two years longer than predicted, you could say, 'It's true the study timeline was put back, and that was because recruitment took longer than expected. However, having a study of this scale means that the results are robust, which is important when it comes to building confidence to treatment approaches in this disease area.' You've answered the question but bridged to the message you want to give, which is that the treatment approach is reliable and trustworthy. Well done you!

Examples of bridging phrases are:

o When we look at the data, what's interesting is …
o If we take a broader view …
o Additionally, what may be of interest is …
o Remember the reason we did the trial is …
o Let me put that into perspective …
o Let me give you an example of how this works in practice …
o Actually, that's not right. Let me explain …
o That may not be the full picture …

Think about your strategy. What's your goal in speaking? Is there a way you can package your responses with some extra content so it brings your audience back to your key message? The added advantage of doing this is that your next question is more likely to be on the subject of your choice, rather than that of your questioner. If you leave yourself outside your chosen topic on the first question, you run the risk that the next will take you further away from it and the following even further. It's like a game of tennis: you want to return to the centre back of the court each time, ready to run the shortest distance possible to the next shot. Don't leave yourself on the edge.

Framing

We talked about framing in the last chapter, when we explored how useful it is for emphasising the context and meaning of your messages. Instead of just stating the bare facts, you also explain what their implications are.

Framing can also be used as a way of handling challenging questions. For instance, if someone asks you what the side-effects of a medicine are, the bare facts might be, 'Nausea, vomiting, diarrhoea and, very rarely, arrhythmia.' However, if your goal is for your audience to feel confident about taking the medicine, it wouldn't be well served by leaving your answer as it is. You would need to add context and frame it. 'Yes, great question. The side-effects are nausea, vomiting, diarrhoea and, very rarely, arrhythmia. What's important to know, however, is that during the trial we didn't have any patients who stopped taking part because of the side-effects, so it's clear that the medicine was well tolerated. And when you put the side-effects into the context of the seriousness of this illness, the patients were very happy to continue with a treatment that may prevent a relapse.' You've given a complete answer but also provided additional, helpful information which frames a statement that could otherwise have been interpreted negatively.

Bridging and framing are two key ways in which you can bring the topic of the question back to where you want it to be. And remember that the place you want to end up is wherever it needs to be to achieve the goal you set at the beginning. Even if you find the question interesting and would love to debate its implications, you need to keep bridging or framing in such a way that you repeat your key messages. Keep your answers simple and succinct, and beware of the curse of knowledge – the temptation to bring in extraneous information that your interviewer could leap on as an excuse to divert the discussion into areas that you'd rather avoid.

How to manage difficult questions

While it's unlikely that all the questions you'll be asked will be overly challenging and intense, you have to be realistic and accept that people don't query what you say for the fun of it but because they want to learn something new. If they're a media interviewer, they're looking for an interesting angle; if they're a fellow scientist at a congress, they may want to scrutinise your work or promote their own; and if they're a panel audience member they may just have an area they'd like to explore. Journalists in particular are questioners to look out for. Their aim is often to rattle you a bit so you say what you 'really' think; they're looking for controversy and drama. The more prepared you are to deal with any type of challenge, the better.

This can seem pretty scary, but there's not a single type of difficult question you can't deal with as long as you've done your homework ahead of time. Let's look at what these questions are.

The personal attack

You may be challenged on why you didn't handle a situation differently, or on a key decision you made. The main points here are to stay calm and to remember your bridging and framing techniques. Dr Anthony Fauci, America's top infectious diseases expert, is a master of refusing to be led into shark-infested waters. For instance, when asked by a news anchor whether he should have advised the public to wear face masks from the beginning of the COVID-19 outbreak rather than later on, he replied:

> I don't regret anything I said then because in the context of the time in which I said it, it was correct. We were told in our task force meetings that we have a serious problem with the lack of PPE and masks for the health providers who are putting themselves in harm's way every day to take care of sick people ... When it became clear that ... the infection could be spread by asymptomatic carriers who don't know they're infected, that made it very clear that we had to strongly recommend masks.[1]

---

[1] www.businessinsider.com/fauci-doesnt-regret-advising-against-masks-early-in-pandemic-2020-7

He framed his answer by describing the situation as it was at the time and, most importantly, spoke calmly and without any defensiveness.

You can also neutralise a personal attack by staying calm and choosing your own positive words rather than repeating the negative language of a questioner. So you might answer the statement, 'So the fact that this trial didn't reach its primary endpoint means that we're no further along with research into this important disease' with 'I agree, the results are disappointing. But let's look at what we've learned. We've moved the science forward, because we know things we didn't before.' Or you might answer the question, 'You must be disappointed by the results?' with, 'Actually, what I find exciting about it is …' Own your language and don't let it, or your tone, be dictated by your questioner.

The controversial question

These are especially important to answer well, as they tend to be the questions that, once on the public record (as all Q&As are), will be tweeted and shared if you don't handle them professionally. Unless there's a controversy that's inherent in your topic, they're likely to be the fourth type of question – the one that's outside your topic and probably also outside your area of expertise. This means that it may be wise to avoid answering the question but without appearing evasive. And to do that you need to put boundaries in place ahead of time that ring-fence what you're prepared to talk about versus what you're not.

For instance, suppose you're asked at the end of a research presentation about the race and gender diversity of the leadership team at the university you represent. If you have a strong view on this and want to make a political statement, you could engage with the question profitably – it's your choice. But if not, make sure that commenting is an active decision and not something you do just because you can't think of what else to say, or because you think you should automatically answer any question asked. If it wouldn't serve your communication goal to stray into this territory, or if it could in fact derail your message, you could reply by setting a boundary: 'In this scientific congress I prefer to stay away from politics and it's not really my area of expertise. My focus is this new treatment approach and I'm happy to take questions about the topic, especially as we have some exciting data here.' You've moved the conversation away from the danger zone and bridged back to your message about your research data. You can see how easy it would be to stumble into an area that could result in your entire study being ignored because of a headline like, 'Top scientist criticises institutionalised sexism and racism in university leadership.'

By the way, never say 'no comment' – that's for politicians and criminal suspects. Always offer something that's moderately satisfying to your questioner, but that also expands on the topic of your choice. And bear in mind that you don't always need to answer a question – sometimes it's appropriate to respond with a message rather than answering it literally. It's not an exam, after all.

Bear in mind that if you're challenged on a weakness in your argument, or something in the science that's inconclusive, you don't need to take the blame for it personally. It may be a flaw that's widely shared across research of this type, in which case you could reframe the question to move it one step away from you. Doctors are often asked about the prices of medicines, for instance, even though it's not their domain. In this case, you could say, 'We know that price is top of mind at the moment. This is where the regulatory authorities, the pharmaceutical companies, and governments need to come together to reach an appropriate agreement. I hope this medicine reaches patients as quickly as possible because of its enormous benefits.' You've broadened the problem into one that's shared by all, not just you, and bridged to your message in the process.

Another version of this question is when it's about someone else's work. When this happens, the last thing you want to do is to speculate or make inappropriate comments by speaking on their behalf. Instead you could say, 'That's a different study, so if you want the detail you'll have to ask the authors. However, what I would say is …' And bring your answer safely back to your own territory.

The question about something highly emotional

When you're working in science or medicine, what you do has a direct impact on people's lives – that's one of the things that makes it so rewarding. But what happens when your work has an unintended and negative consequence, especially if it plays out on a personal level? If you dive in with a scientific answer to an emotive question, it won't go down well. Suppose there are suddenly reports about a number of deaths associated with a medical procedure you've been involved in trialling, and you're asked about what might have happened and how you feel about it.

Here you follow a three-part answer:

1. Find a zone of empathetic agreement – an area of non-controversy with which everyone will align.
2. Provide your own perspective.
3. Give a forward-looking statement that moves everyone forward productively.

In this case, your answer could run like this:

1. 'I was deeply saddened when I heard about these reports. They're very concerning, and naturally my heart goes out to the families of everyone involved. Any death is deeply upsetting.' This establishes a connection with your questioner because it's something that can't be contested. It also provides a basis of agreement that encourages them to listen to what you have to say next.
2. 'At the moment we don't know exactly what the situation is, and if there's a link between the procedure and these outcomes, because we're still carrying out an investigation. It's too early right now to say definitively what happened, but it's our mission to find out as soon as possible.' This gives your take on the situation.
3. 'So we're looking at these cases in detail to see what we can learn from them. It will be our top priority. As soon as we have more data, we're committed to sharing it straight away, because the most important thing is that people undergoing this procedure have good-quality information. In the meantime, people should ...' You've said what you're going to do next and provided a bridge to one of your messages. You've also moved from the personal to the general, thereby avoiding speaking about individual cases. And you haven't speculated about what could be the cause before you have all the facts.

The hypothetical question

This can be a hard one for scientists to resist, because so much of what you do involves coming up with theories and hypotheses and then proving or disproving them. It's what you love. However, my advice is to be very careful when engaging with a hypothetical question when you're in front of audiences outside your specialty area. Hypothesising in these situations is speculation, and it is prone to being misinterpreted by those who are less familiar with the topic. Headlines are full of dramatic exaggerations based on conjecture: 'Thousands of Americans *could* be at risk from rare water contaminant'; 'Chocolate ingredient *may* be linked with cancer.' Ask yourself whether hypothesising would help your audience focus on your message or detract from it. If you feel it would help, go ahead and answer it but make sure it's an active choice. If you think it wouldn't, you can use the same deflection and bridging techniques as for the controversial question. Or you can simply say, 'As a scientist, right now it's not helpful to speculate. But what we do know is …'

Prepare, prepare, prepare

If, while you were reading the above, you found yourself thinking that you'd struggle to do any of this on the spur of the moment without freezing, becoming defensive or digging yourself into a hole, you're right. That's why you need to predict the questions you'll be asked ahead of time and have answers prepared. That way you'll become just as great a communicator and connector as when giving your talk.

To prepare, your first task is to think about the questions you might be asked. This isn't as hard as you might assume – I believe you can identify at least 95% of them by employing your common sense and doing a bit of research. Become an investigator: read around your subject and search for what other people – especially your critics – say about it. What's the government policy? How much do things cost? What are the implications of your work? What 'angles' do the media usually take when this topic is raised? There will always be the final 5% of questions that are wild cards, but once you feel confident with the techniques I've taught you, you can deal with those if they happen.

Next ask someone to help you prepare, as they can put themselves into the questioner's chair more easily than you. Give them your list of questions and suggest they throw in a few curve balls of their own. You'd be surprised at what they come up with. I often have clients who say to me, 'I'll never be asked that', to which my reply is, 'Well, I found it on the Internet.' It's far better to be over-prepared and surprised by the ease of the questions than to be blindsided by a couple of nasty ones.

The benefit of preparing is that it not only allows you to handle the Q&A professionally, but also relieves you of some of your mental load. Giving a talk and then taking questions is exhausting, as is giving an interview, and when you're tired you're liable to make bad decisions. That's why you need to decide what you're going to say ahead of time, and rehearse it again and again, if you don't want to make a bad judgement call on the day. If you find yourself thinking original thoughts in the middle of a Q&A, be wary – this is not good. You're in danger of swimming with the sharks.

One final tip

If there's a controversial area of your talk, or one you know you're bound to be asked about, a clever way of using the Q&A is to avoid covering it in your talk as you know full well that it's the first question you'll be asked. There are a couple of reasons for this. First, it means you have a predictable question that you can prepare for and that will use up some of the question time, thereby leaving you with less fortune-telling to do. Second, it allows you to address the issue in a top-line way rather than having to answer a more detailed and challenging question about it later.

Let's look at an example so you can see what I mean. Suppose you led a massive study showing that Drug A is effective in a certain area. However, you've learned that another study has just come out showing that Drug B is also good. Rather than comparing them in your talk, which would have the result of partially neutralising your positive messages, you would leave it out altogether. When you're asked a question about Drug B, you'll have a well-prepared answer that covers the differences between the two modes of action and trials, and why comparison is not possible without a head-to-head comparator trial. Craig Venter used this technique to great effect when he presented his creation of the synthetic cell. He only briefly mentioned liaising with ethicists in his announcement, but he clearly knew it was the first thing he'd be asked about as he had a slide ready to explain it during the Q&A. You don't need to go so far as having a slide, of course, but try to rein in your scientific instinct to be thorough and explain everything up front – keep something in reserve.

The main points

o It's just as important to prepare for the Q&A as it is to prepare for your talk, especially as formats such as media interviews and panels are mostly Q&A based.
o Questions can range from easy to hard, but they all need answering in such a way that will bring your audience back to your messages.
o Various techniques can be used to handle questions professionally and use them to achieve your goals.
o The more work you do ahead of time to prepare, the more successful you'll be.

Chapter 6
Communicating numbers, risk and uncertainty

> *An unsophisticated forecaster uses statistics as a drunken man uses lampposts – for support rather than illumination.*
>
> *Andrew Lang, Scottish poet*

It would be an unusual talk by a scientist or physician that didn't include statistics, risk factors or numbers of some kind. Given that your task is to win over people to your way of thinking, it's natural that you'd want to back up your points with facts and figures; it's *logos* again, and an important element of scientific speaking. Yet, particularly when speaking to people who aren't your scientific peers, it's also the area that can most easily land you in hot water if you're not careful. This is important to consider because big things are at stake here – people may make decisions based on what you say and they should be doing it from a well-informed perspective.

Simplicity is key when you're talking about risks and statistics. It's a myth that all scientists are proficient with numbers and it's certainly not true of the general public; recent research showed that only one in five adults could today scrape a pass in GCSE maths, the exam taken by UK school students at 16 years of age.[1] There are ways of putting across your information with

---

[1] www.kcl.ac.uk/news/research-reveals-how-poor-maths-skills-are-holding-the-uk-back

clarity and persuasiveness, and there are ways of doing it so it confuses and misleads your audience. This chapter will show you how to achieve the former and avoid the latter.

Before I go on, I want to make clear that there are whole books written about the communication of numbers, risk and uncertainty – this single chapter is certainly not attempting to be the definitive guide to the subject. What I'm doing here is to share the main insights I've gained in my 25-year career in scientific communications. I'll start by examining the inherent difficulties in talking about risk, and how to do it effectively, then I'll tackle the topic of numbers and statistics. For a list of additional resources for communicating risk, please see https://scientificallyspeaking.com/risk-communications-resource.

The challenges of communicating risk

Talking about risk within science is fraught with difficulties, mainly because so much depends on your audience's personal perspective. What's reassuring to one person can be downright terrifying to another, and what can be easy to accept for a scientist can be confusing for a non-scientist. This means that *how* you communicate risk is as important as *what* you communicate because you have to take care not to be misunderstood by the specific audience you're addressing.

Take the public health catastrophe that was dubbed the 'pill scare' back in October 1995. The UK's Committee on Safety of Medicines sent a 'Dear Doctor' warning letter to British doctors concerning the use of third-generation combined oral contraceptive pills.[2] This warning was based on at the time unpublished studies suggesting that a risk of deep venous thromboembolism in women taking the pills increased compared with women on second-generation combined oral contraceptives. The statement was problematic because it didn't put the risk into context. The relative risk in the studies suggested an increase of 100% – that's a twofold increase. But the *absolute* risk for women went from one in every 7000 to two in every 7000.

As anyone in possession of the facts could see, what the figures effectively meant was that the risk went from very small to slightly less small. However,

---

[2] UK Committee on Safety of Medicines, 1995

no media planning was done and the way it was reported in the media was that the risk of thrombosis 'doubled' for women on the pill. This naturally led a significant number to switch brands or stop contraception altogether, with a subsequent increase in the number of unintended pregnancies. Even leaving aside the stress and anxiety caused, and the time taken by GPs to support countless worried patients, the resulting cost to the NHS of the increases in births and terminations was estimated at £57 million.[3] This was all down to the fact that media consumers were given no perspective or context to help them make an informed decision.

So let's look in more detail at why communicating risk is such a minefield, and how this manifests itself in speaking situations (I'll go into what you can do to solve the problem in a moment).

There's a difference between actual and perceived risk

Each of us has different experiences of life and varying ways of looking at things. We instinctively understand that a person who's undergone a traumatic but rare experience, such as a serious car accident, might feel scared about it happening again. There's no more logical reason for them to be wary of using the roads than there is for anyone else, but they don't see it that way and nor would we.

This is because we humans don't think about risk in a mathematical way. We take part in dangerous sports such as rock climbing or paragliding, which make no sense when you look at the potentially negative outcomes, yet we can also feel too scared to get on a plane. Our stomachs churn when we take a glass-floored elevator even though common sense tells us it must be safe. And we worry about catching a rare but potentially fatal virus like Ebola while at the same time living the kind of unhealthy lifestyle that's far more likely to lead to our early demise. This illogical thinking became apparent to me when I worked in the field of toxic shock syndrome and discovered that many young women were terrified of contracting it, even though only a handful of cases occurred each year. How we perceive risk lies in how risky it feels to us, rather than how risky it actually is.

The difference between actual and perceived risk is also illustrated by the fact that we tend to view certain things as being riskier than others, based on the

---

[3] https://pubmed.ncbi.nlm.nih.gov/10652971

emotional characteristics of the risk itself. Have you ever asked yourself why topics such as GM food, chlorinated chicken, BSE, toxic shock syndrome, Ebola, acid rain and MMR have provoked such a lot of debate? Wonder no more. Peter Sandman, a retired risk communication consultant, is an expert on risk perception and carried out valuable research into it in the 1980s that still holds true today. The result of this research was a list of 12 factors that explain why something might be perceived as especially risky (he describes the inflammatory nature of the risk as provoking 'outrage'). This list is, I believe, even more pertinent today given the proliferation of scare stories on social media, and goes some way towards explaining why they're shared so widely and with such speed.

Here is his overview of 'The 12 Principle Outrage Components'. Notice how emotional the factors are.

1. *voluntary or coerced* (people cope better with risks over which they have a choice)
2. *natural or industrial* (people are readier to accept an act of God than a human-made threat)
3. *familiar or exotic* (people are less scared of what they're used to)
4. *not memorable or memorable* (there's always more outrage when there are images, slogans, or nicknames to latch onto)
5. *not dreaded or dreaded* (disgust or fear can provoke outrage)
6. *chronic or catastrophic* (a sudden threat is harder to cope with than one that people have had time to get used to)
7. *knowable or unknowable* (the unknown is more frightening than what's understood)
8. *controlled by oneself or by others* (when people can exert control over a risk they feel less fear)
9. *fair or unfair* (people take the risk but someone else gains the benefit)
10. *morally irrelevant or morally relevant* (people can identify a moral issue)
11. *trustworthy or untrustworthy* (the source of the risk is deceitful or dishonest)
12. *responsive or unresponsive* (people are excluded from the decision-making process).[4]

---

[4] 'Responding to Community Outrage: Strategies for effective risk communication', www.psandman.com/media/RespondingtoCommunityOutrage.pdf. I have found that www.psandman.com is a valuable resource when planning communications for controversial issues.

Each risk can be described as having either the first side of the characteristic or the second, and the greater the number of second characteristics (or outrage factors), the more of a challenge your communication will be.

Let's use an example to show what Sandman was getting at: the potential import of chlorinated chicken from the United States to the United Kingdom, a subject that has been the subject of much heated debate. What is it about this that provokes outrage? Moving down the list, we can see that the reasons are many, as our example points to 10 of the 12 outrage factors: it's coerced (it's perceived as being forced onto the UK by the US as part of trade negotiations); industrial (human-made); exotic (from abroad); memorable (images of chickens are all over the media); dreaded (it seems repellant to add chemicals to a natural foodstuff); controlled by others (originating from the United States not the United Kingdom); unfair (members of the public risk their health by eating it and UK farmers are undercut on price, but the United States gains the benefit); morally relevant (from both a public health and animal welfare perspective); untrustworthy (many people don't believe government assurances about food health after the BSE scandal); and unresponsive (consumers aren't able to directly influence the decision to allow it).

I don't know about you, but if I were a food scientist giving a talk on why chlorinated chicken isn't such a bad thing, I'd want to think very carefully about how I talked about the risks. There would be little point in me trotting out a bunch of statistics showing how few people had been harmed by it or the benefits of cheap chicken – that would be like King Canute trying to stop the tide. I'd only be working against human nature, because it's not possible to fight emotion with logic and still win.

People aren't calculators

Numbers appear to be definite, yet it's the seemingly concrete nature of numbers that can create misconceptions. For instance, if I read a movie review I'm aware that it's the reviewer's subjective opinion, but the star ratings for the film – although just as subjective – seem more objective and credible to me because they're expressed as figures. In the same way, the general public can assume that science is science and that all studies are created equal, whereas scientists know that this is often far from the truth and there can actually be a significant difference between the quality of various research pieces. There's often a story behind the numbers that creates nuance, and this can be hard to appreciate for many; the numbers are black and white – but only in the context of the overall study.

Yet this context is not always made clear by scientists. Are the risks of taking a new drug based on early- or late-stage research? Are they based on a handful of triallists or 200, or even 200,000? I once talked to some researchers who were exploring the effectiveness of a potential new treatment for glioblastoma. They showed me the 'before and after' brain scans, which were astounding – in the 'after' scans, it looked as though the tumours had disappeared. However, on further discussion I discovered that the scans were in fact those of mice and that it was a pre-clinical trial. Context is critical to how we interpret information.

The trade-off factor

The notion of trade-offs is embedded within risk perception: is taking the risk 'worth it'? It's a subjective position. And what would be the results of *not* taking the risk – might something even worse happen? We make this kind of trade-off all the time, mostly without realising it. For instance, someone might decide it was too risky to get on a plane to go on holiday during the COVID-19 pandemic, but if one of their family were in trouble abroad they would have no hesitation in doing so. The trade-off has changed because the context has changed.

Gaining vs losing

Research shows that people are more likely to take a risk to avoid a perceived loss than to pursue a possible gain[5]. So when speaking about trade-offs, it's also important to consider how you frame the risk in terms of what is lost or gained. Saying 'these results show that the vaccine is effective in 90 out of 100 people' is positive framing, but the same results can be framed negatively by saying 'the vaccine was not effective in 10% of people'.

Crucially, the motivational impact of a possible loss or gain will still depend on an individual's life experience and personal perspective. So choose your framing in terms of both your goal and the likely mindset of your audience. Sometimes, to illustrate the point you are trying to make, it could be wise to

---

[5] https://www.apa.org/science/about/psa/2015/01/gains-losses

give both a negative and positive frame – for example, 'While the vaccine did not protect 10 people in every 100, the vast majority of those who took it – 90 in every 100 – did not go on to contract the virus.'

Science isn't absolute

Finally, people like absolutes because they make them feel safe, but unfortunately science rarely delivers them (at least in the fields of biology and medicine). New discoveries are constantly being made, which means advice changes and experts sometimes disagree with one another. To you, this probably seems normal and unavoidable, but to non-experts it can be problematic because the burden of decision-making now falls to them. Whose opinion should they trust? And why?

As you can see, navigating the choppy waters of communicating risk and uncertainty is a hazardous activity in its own right. It's tempting to blame the media for misreporting the nature of risk, but you have to accept that journalists aren't there to provide a scientific information service. When they write a headline they're communicating their version of the risk – the one that will interest their readers and viewers – rather than the scientific side. When speaking to an audience outside of your peers, it's your job to talk about risk in a way that both members of the public and journalists can understand without them feeling panicked or overwhelmed. How you do that is what we'll explore next.

How to talk about risk

How do you communicate risk and uncertainty and still achieve the goals you set for yourself when you planned your talk? There are a number of steps to take.

Go back to basics

Think about your objective. What do you want to happen as a result of you communicating the risk? Do you want people to change their behaviour, worry about something, not worry about something, start doing something or stop doing it? What do you want them to feel – hope, fear, excitement? And what is it realistic to expect them to do? For instance, you may want to influence those in government about climate policy, or to ask the general public to be more thoughtful about climate change and alter their behaviour.

Consider your audience

Who are you speaking to? The general public? A group of patients? Policy-makers? Step into their shoes. What are they likely to be thinking about the risk? How can you make your message about it relevant to them, and explain it within their frame of reference?

Predict the outrage

When I worked in risk management in pharmaceutical communications, my team and I would frequently be asked to advise on how to launch medicines, and in particular how the side-effects and other risky aspects of them should be spoken about. The first thing I'd do would be to use Peter Sandman's risk-evaluation chart to plot out the areas of concern. If I identified more than three or four outrage factors, I knew the topic was highly sensitive and that we'd have to be extremely careful and precise about how we communicated the risks. I suggest you do the same because the last thing you want is to provoke a media scare. If you find yourself with a long list of concerns, you'll need to spend time thinking carefully about how you'll talk about them.

Use the right comparisons

Sometimes it can be helpful to frame the risk by providing a comparison that guides people towards the conclusion you want them to reach. For instance, if you wanted to recruit volunteers for a clinical trial, you might liken the amount of time they'd need to spend in the clinic as being no longer than it takes to drink a cup of tea. We all know what that's like and how long it takes, and it's a non-threatening comparison.

However, tread with care. You need to know your audience well because it's easy to create a comparison that works on a scientific level but not on the level they're thinking on. Don't be like the British government advisor who told parents that the risk of their child dying by catching COVID-19 at school was lower than the risk of them dying in a road traffic accident on the way there. It made sense on a logical, scientific level, but it offended many parents who felt it was inaccurate and unfeeling (and still risky). Remember that you can't fight emotion with logic: you have to find an emotional way in to connect with your audience's thinking that makes sense to them.

Don't dodge the risk

It's tempting to skirt around the uncertainties because you know your audience may feel uncomfortable about them, but – depending on your goal – this is not always a good idea. Similarly, avoid talking in absolute terms because science is rarely absolute. Instead you could say, 'This is as close to a confirmation as we have' or 'Until the data prove otherwise, this is what we know'. Also be careful with the word 'safe'. People usually interpret that as meaning 'no risk', when what you probably mean is 'minimal risk'. Instead, it can be helpful to talk about the risk in terms of balancing it out with the benefits, thereby making a decision in the context of the whole picture rather than in isolation.

Communicating numbers and statistics

In talking about risks and uncertainties, you'll inevitably make use of numbers and statistics. However, just like risks, numbers can be misinterpreted if you're not careful. This can lead to misunderstandings and – in the worst case – serious unintended consequences, so here are some ways to make sure that you leave your audience with the right impression.

Give your numbers a context

For most audiences, numbers on their own don't mean anything, nor are they especially memorable or interesting. To give them a meaning, you need to present them within an expert context and add your own opinion. This is especially important if your scientific communications are based on studies and data, when your numbers are presented in the context of a particular piece of research. Whether you're on an expert panel, advising on policy, giving a media interview, speaking at an event or providing evidence as an expert witness, you have to explain what the backdrop is if you're not to mislead people.

Related to this, remember that statistics are only tools that ask questions of data, and that they're therefore only relevant to the data they're looking at. They're not facts in themselves. You need to state the statistic and then give your interpretation, which provides the context. That way you're not just saying what the number is: you're explaining what it means. This is especially true when you're communicating p-values and risk factors. Here's a breakdown of the best way to put across your meaning for these:

For p-values

- *State the statistic or function:* 'We looked at the p-value to see if these results were significant. Could they have happened by chance?'
- *Share the result:* 'The data show a p-value of 0.0254 …'
- *Give your interpretation:* '… which means that there's only a 2.54% chance of this happening randomly or by accident.'
- *State your expert opinion:* 'This proves that the compound is showing promise in such a new field.'
- *Give the context or advice (the 'so what' for your audience):* 'This is exciting, but the research has to be validated before it becomes a new treatment option.'

For risk factors

- *State the statistic or function:* 'We looked at the absolute and relative risk reduction between the treatment arm and the control or placebo arm.'
- *Share the result:* 'At the end of the study in the control arm, 70% of patients had died compared with 35% of the treatment arm. So the absolute risk reduction is 35%, and relative risk reduction for the therapy is 50%. This means that if we were comparing two groups of 100 patients, 35 more patients would still be alive in the treatment group at the end of the study.'
- *Give your interpretation:* 'This means that the death rate in the treatment group was half of that in the control group …'
- *State your expert opinion:* '… which shows that the drug has a significant impact on the survival of patients.'
- *Give the context or advice (the 'so what' for your audience):* 'Given these data, we look forward to this treatment becoming available soon.'

For confidence intervals

o   *State the statistic or function:* 'We looked at the confidence interval to measure how sure we can be that the estimates obtained in the study were accurate.'
o   *Share the result:* 'At the end of the study the medicine's efficacy was 26% with a 95% confidence interval of 15–44%.'
o   *Give your interpretation:* 'Although the study suggests that 26 people in 100 would benefit from the medicines …'
o   *State your expert opinion:* '… the results were not very precise and the benefit may be as wide as 15 to 44 people in 100 benefiting.'
o   *Give the context or advice (the 'so what' for your audience):* 'We are looking forward to seeing more data in this area before we can be more definitive.'

Make your numbers easy to digest

Very small and very large numbers are difficult for people to grasp. For example, 0.00001, or billions and trillions, are too extreme to have relevance because we so rarely use them in normal life. There are various ways of getting around this:

o   Whole numbers are easier to understand than decimals, so say 'one in a million' rather than 0.000001%.
o   When you have to use extreme numbers, package them up into understandable units of time or space. For instance, a million seconds is 12 days; a billion seconds is 30 years; a trillion is 30,000 years. One in 1000 is one child in a secondary school; one in a million is one person in a mid-size city such as Birmingham in the United Kingdom or Austin, Texas in the United States.
o   Use comparisons that help your audience to make sense of the numbers. An article[6] about climate change, headlined 'Earth loses 28 trillion tonnes of ice in less than 30 years', made the enormous figure easier to relate to by saying: 'To put the losses we've already experienced into context, 28 trillion tonnes of ice would cover the entire surface of the UK with a sheet of frozen water that is 100 metres thick. It's just mind-blowing.' However, always make sure your analogy is something your audience will understand and appreciate; it has to be something that they can anchor to, which is concrete and real to them.

---

[6] www.theguardian.com/environment/2020/aug/23/earth-lost-28-trillion-tonnes-ice-30-years-global-warming

If you want to minimise the impact of a statistic

Suppose you're asked to comment on reports about a (fictional) medicine in wide use that's reportedly been blamed for the hair loss of a celebrity. It's a noted and rare side-effect of the medicine, but although it was only one person who was affected it's the sort of story that could end up persuading patients to stop taking it. You could depersonalise this situation by putting it into perspective and using a comparison: 'We know this can happen, but let's put it into perspective. It's extremely rare, and by that I mean it's one in 100,000. That's the equivalent of one person in Wembley Stadium.'

Another way of minimising the impact of a number is to break the rule that says you shouldn't use a very small percentage. 'While this is obviously a worrying report, it's rare. The proportion of patients who had this particular side-effect was 0.00001%.' The repetition of the noughts emphasises the small size of the figure.

If you want to maximise the impact of a statistic

Your audience is more likely to latch onto numbers that represent a human face than those that are abstract concepts. This means that if you describe a statistic in terms of people, you'll draw attention to it more fully. For instance, if you want to encourage donations to a cancer charity, 'One in two of us will experience cancer in our lifetimes' is more personal than '50% of the population'. Or if you want people to avoid social contact as COVID-19 numbers increase, you can explain the reproduction rate (R number) in terms of people: 'The R number is currently 1.8, which means that 10 people who have COVID-19 will infect a further 18 people.'

This would be a good addendum to the hair loss case-study. You could also ensure that the small risk of hair loss is balanced with the higher risk of the illness for which the medicine is prescribed: 'And this needs to be put into context. Remember that this medicine is used in people who have high blood pressure. High blood pressure increases the chances of heart attacks and strokes, and we know that more than one in two people in America will develop heart disease in their lifetime.'

Explain your sample size

Did you know that 25% of The Beatles were murdered? Obviously this was only one person, but you can see how a high percentage number and small overall sample size affect the meaning of the percentage. You need to put your statistics into context. For instance, in Switzerland, where I live, our COVID-19 infection numbers on one day went as low as nine people and then doubled to 18; 'doubling' sounded like a lot, but in a population of seven million it was almost meaningless.

Also, think about the value of your long-tail figures, which might tell an interesting story in their own right. If you worked on a trial which had a minimal impact on most patients but provided a cure for eight of them, you could talk about how this has given you the basis on which to do more work in the light of a possible genetic marker. Sometimes the story lies not in the top-level data but in the outliers.

Leave the statistics out

Depending on the nature of your talk and of your audience, it might not be necessary to use numbers and statistics at all. Why not go straight to talking about the implications of the data instead? And if you do use them, make sure they're a support to your main points and stories, rather than the lead-in. Numbers on their own are rarely interesting enough to be put up in lights.

You can see that the way you talk about your numbers, risks and statistics should be guided by the effect you want to have and the people to whom you're speaking. There's a huge amount you can do to make yourself understood simply by thinking carefully about how you present and explain the data.

The main points

- Numbers, risks and statistics are valuable tools for backing up your points, but if communicated badly can create enormous problems for you.
- It's your responsibility to talk about them in a way that doesn't mislead or confuse people.
- We perceive risk in an emotional, rather than logical, way.
- Plan how you talk about risk and statistics rather than assuming that the rational, scientific approach is best.

Part 3

How to speak engagingly in any type of situation

Chapter 7
Energy

All things are ready, if our mind be so.

William Shakespeare, Henry V

I'd like you to think about a speaker or presenter you love watching. Someone who inspires you. They don't have to be a scientist or doctor – they could be anyone you've seen on stage or screen who always holds your attention to the end. Ask yourself what it is that makes their performance impressive. Is it what they say? The way they dress? Their visual aids? The way they move?

Whenever I ask my clients to do this, they all end up saying the same thing, so I'm going to take the liberty of making some assumptions about the answer you'd give: 'They're enthusiastic about their topic! They're passionate! They make me care about what they're saying!'

When I ask how the speaker does this, they say, 'They show their enthusiasm in how they speak.' And what does that mean, I ask? This elicits some more specific answers: 'They smile. They speak with energy. They stand up tall. They look at the audience. They relate to me.' In other words, they bring their energy to the topic and appear physically comfortable with being on stage or on camera. They speak *to* people rather than over their heads. It's rare that I receive the answer, 'Oh, he's so funny,' or 'Her data are incredible.' That's not what people remember – it's the passion and the emotional connection that make a deep impression.

To establish this connection, you need to have energy – the vitality and passion that you bring to your talk. Because if you want to win people's attention and make them interested in and excited about what you have to say, you first have to show them that *you're* interested and excited. You, as the messenger in the communication model (see Figure 9), have an impact on the success of any communication and your energy is a critical component of this. If you're not enthusiastic there's no chance that anyone else will be.

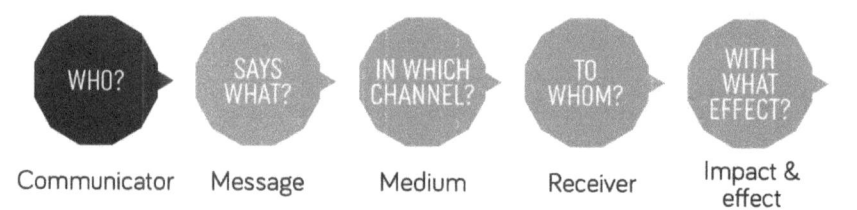

WHO?	SAYS WHAT?	IN WHICH CHANNEL?	TO WHOM?	WITH WHAT EFFECT?
Communicator	Message	Medium	Receiver	Impact & effect

Figure 9: How you come across as a communicator is critical in successful communication

Speaking with energy makes your communication engaging and effective – people will sit up and pay attention when you come on stage and they'll remember what you've said. If you think about the speaker you called to mind earlier, you probably find their enthusiasm infectious and their sense of presence inspiring. I wouldn't be surprised if you listen to them with complete absorption, forgetting about your everyday distractions and focusing all your mental energy on them. That's because they're fully present on stage, keeping their audience top of mind so that they can respond instinctively to shifts in mood by adjusting their mode of delivery and even their content if needed. That's how they win the war of attention.

How to speak with energy

You may be wondering how on earth you can feel energised and passionate when you're shaking with nerves as you walk onto the stage, or starting a Zoom presentation to hundreds of unseen people. This is an understandable concern, but it's important to acknowledge that speaking in public should never feel normal; if it does, something is wrong. You don't need to be paralysed by anxiety – that's never a good thing – but you should feel as if what you're about to do is a big deal because it sharpens your focus and gives you an edge. You can see nervousness as a power booster for your energy. Nerves are a signal that you care about this situation enough to be a little anxious; you should pay attention and prepare accordingly.

Having said that, you want to enjoy the experience as much as possible, if for no other reason than that your audience will be more likely to enjoy it too. So how do you put yourself into the right frame of mind for an energised performance? The solution lies in the way you prepare for the experience and how you deliver when you're in the middle of it. In fact, when you're on stage or on camera, I believe you need to generate at least 20% more energy than you would when talking to people one to one, so it's worth knowing how to get yourself into the right frame of mind.

I'll cover all aspects of how you can get your mind and body ready for speaking in the next chapter, but for now I'll touch on some key ways of increasing your energy and vitality.

Connect with your purpose

First and most importantly, to convey your passion for your subject you need to come to your presentation or interview ready to deliver on your goal, and this involves connecting with your purpose. Why are you here? What is it you want to achieve? What do you want people to go away thinking, feeling, and ready to do?

Of course, it can be challenging if you're nervous about the situation, so it's helpful, before you begin, to say to yourself, 'I know my stuff. It's valuable to other people. And what I want for them is …' Remember that you're doing your audience a valuable service by sharing your expertise with them, and if you come across as confident and enthusiastic, then you'll help them to care about it as much as you do.

Prepare ahead of time

Second, it's important to reduce your mental load so you will have the head space to connect with your audience when you speak. Presenting in public, whether on stage or on camera, can be stressful. There's a huge amount going on and a lot to think about. If you're working out what your message is and worrying about what to do with your hands and where to look, it can be hard to focus on your audience at the same time. That's why preparation is critical. The more you practise, the easier it is for you to show your passion and connect. Instead of worrying about what you're going to say and how you'll handle the Q&A, you'll have practised your content so many times that it doesn't seem too daunting to deliver it with confidence.

Consider your mental travel time

Third, turn off your phone for the hour before you speak. Being present and mindful is an essential component of vitality, and to gain other people's attention you have to be a master of your own. Your talk should have your full priority in the run-up to going on stage or sitting in the interview chair, and the last thing you need is to tap into your emails 10 minutes beforehand only to learn that your paper submission has received extensive peer-review comments that will take hours to address. You'll have half of your brain dealing with that when you should have all of it focusing on your goal and your audience. That's why, when I'm with a client who's about to present, I either ask them to turn off their phone or I take it away from them for the duration. To create an impact, you have to generate a powerful presence, and for that you have to *be* present.

Watch your tone and body language

Professor Albert Mehrabian, Professor Emeritus of Psychology at the University of California, conducted studies showing that only 7% of what we say makes an impression on our listeners. Some 38% is tone of voice and 55% is persona (how you look and your body language). For me, the most important thing about this research is not that what you say isn't important, but that the way you say it has to reinforce its meaning. If you want your words to have an impact, your vocal tone and body language must be aligned with what you say. Congruence is key.

If your first sentence is, 'I'm so pleased to be here, and thank you to the organisers for having me', but spoken in a monotone, people won't think you really are pleased to be there – they'll assume that you'd rather be anywhere but on that stage. You can lose people at hello as much as you can win them, whereas if you smile warmly, look at the audience and have an energetic tone of optimism, you're congruent and instantly engaging. A helpful tip is to smile before you begin any talk because it elevates your mood and encourages your audience to connect with you. And make sure you hit your opening words crisply and with volume.

Move on swiftly from a mistake

Everyone gets things wrong sometimes, and it's not unusual to fluff a line or forget to mention something, especially when you're under pressure. If you go blank, simply refer to your notes. And if you make a mistake, you can just apologise and carry on: 'Sorry, I lost my train of thought there. Where was I? Oh yes …' Or, 'There's so much data here that I forgot one of the most important points. I should have said earlier …' It's helpful to have a couple of stock phrases to hand for these kinds of situations so you don't have to think about what to say.

In my experience, if you make a mistake it will probably be the one thing you remember about your talk. However, I can guarantee that if you move on swiftly and put energy into the rest of the talk, your audience will have forgotten about it by the time the presentation ends. They're remembering the piece in its entirety, whereas you're focusing on just one element of it. So if this happens to you, move on and make your next point a strong one – hit the ball out of the park. If you finish well, your audience will recall the ending far more strongly than the one phrase or sentence that was inelegant.

I hope you can see how essential energy is for engaging your audience and helping people to connect with your messages. It's why you're speaking instead of writing – you're wanting to carry people along with you, rather than treating them as passive recipients of your information. In the next chapter, I'll explore various ways in which you can get yourself mentally and physically ready to speak with impact.

The main points

o   Energy is usually the missing ingredient when speakers fail to connect with their audiences.
o   You can cultivate it by focusing on your purpose, preparing ahead of time, paying attention to your physical presence and not letting mistakes knock you off course.

Chapter 8
Managing your nerves and speaking with confidence

The human brain starts working the moment you are born and never stops until you stand up to speak in public.

George Jessel, *American actor*

Almost everyone feels nervous when they're about to speak in public, whether it be on stage or under the lights of a TV studio. I've been speaking professionally for many years and I still have a churning stomach before I begin. When you feel stressed, cortisol and adrenaline course through your body, triggering the 'fight, flight or freeze' response. This manifests itself in feelings of tension and agitation, shortness of breath, a dry mouth and other 'symptoms' that aren't so helpful if you want to speak with confidence and energy. These hormones also reduce your ability to reason and empathise with your audience, which can make listening carefully to questions and responding with emotional intelligence a challenge.

Fortunately, there are lots of ways to calm yourself down and give yourself the positive energy you need to deliver a powerful performance, and these are what this chapter focuses on. You'll learn how to prepare mentally and physically for what could be a career-changing opportunity – you may even enjoy it!

Practice and rehearsal

One of the keys to performing well is to have practised and rehearsed ahead of time. This reduces your mental load so that instead of worrying about making mistakes or wondering what to say next, you can free up your energy and focus on engaging with your audience. Think of yourself as a TV chef who's done the prepping and chopping ahead of time, so that when the cameras roll all you have to do is entertain your viewers by demonstrating your delicious recipe.

Opening and closing

When you speak, I recommend focusing on ensuring that you have a strong beginning and end, because those are the parts that make the most impact on your audience: the beginning because it grabs their attention, sets the tone for what's to come and increases the chances that the audience will stay the distance; and the end because, as the final thought, it's often what's remembered most.

Preparing your opening is also important because, as you walk onto the stage or put yourself in front of the camera, it's a huge comfort to know that your first few sentences are ready to trip off your tongue. If you know what you're planning to say, you can concentrate on looking at your audience and delivering your words with energy.

Be ready to break the mould to grab attention. The standard opening line for a conference presentation is often something like, 'I'm so grateful to be here, and it's an honour to speak to you today.' That's fine, and uncontroversial, but it's not memorable. It's much more interesting to start with something more original, such as:

- As I was preparing for this talk, three things struck me.
- A lot of people don't know …
- Whenever I talk about this topic I find that the biggest misunderstanding is …
- Today is the anniversary of [insert appropriate event] and …

A surprising or attention-grabbing opening perks up your audience and brings them onto your side, which in turn helps you to deliver with confidence. Endings are equally important because it's common to see speakers petering out at the end of a presentation rather than finishing with a bang. You want people to go away feeling impressed.

I suggest that you script both your opening and closing sentences word for word. In fact, scripting the first line of every slide can also be a good idea, although it's up to you how prescriptive you are. Whatever you do, the point is to have a plan and to physically practise saying the words aloud so they sink into your muscle memory.

Practice

Practice, which you do on your own, comes before rehearsal, which is what you do in a setting that's more akin to the real thing. You should start practising well in advance of your talk or interview, with regular repetition. The reason why practising regularly is important is because in between sessions your brain continues to work on it without you realising, using its default mode network – it's as if you get extra practice for free. When I was a spokesperson for a medical company, I would practise my media interview phrases and answers in the car to and from work, saying them out loud. If I had a set of interviews the next day, the night before would see me speaking them to myself on the way home; the more I said them, the more confident I'd feel and the better my delivery would be.

Practising has the obvious benefit of reducing your mental load on the day, but it also helps you refine your message and deliver a fluent and compelling talk. This takes time. To become verbally proficient takes sustained effort; it may look easy when you watch television presenters or experienced speakers, but to explain complex topics in a way that your audience understands takes experience. Most people who do this well have spent many hours talking out loud about their specialty, and it shows. Their phrasing is succinct, and they have a better connection with their audiences than those who haven't put in the time. They're also more likely to risk telling a joke or engaging in some two-way interaction, because they have the mental space to be in tune with the people watching them (and with themselves). I'd also be willing to bet that they're more likely to enjoy the experience.

So how do you practise? First, set aside enough time before the big day. For conferences this may be weeks, whereas with media interviews timelines are usually shorter. As a rule of thumb, you should spend about half the time available to you on developing your content and half on practising and rehearsal. The more important the event is to you, the more time you should spend on the latter because the quality of your delivery is what will make the most impact.

Next, create your slides (if you're using them) and start speaking them aloud as you sit in front of your computer screen. You're not worrying about being polished at this stage, only about making sure that your points are clear and engaging, and that you can talk through them without any problems. Once you feel familiar with the material, switch to stand-up mode (if that's the way you'll be presenting). It's a good idea to record yourself on your computer or phone so you can see how you come across. You probably won't want to do it and may be tempted to skip this step, but if you're serious about delivering a great presentation, do it. You can also use the recording to ask people for feedback: 'How do I come across? What's the key message you took out? What was good? What could I improve?'

In all this, be kind to yourself and keep your self-feedback as positive as possible. For instance, 'Stop looking away from the camera' could be, 'I could look more at the camera.' And if, like most people, you hate watching yourself on film, just say, 'I'm not going to go there. I'm just going to look for whatever is good. I'm smiling. I'm standing up well. I see an expert who's professional and with a wealth of knowledge.' Give yourself five positives for every negative.

Above all, at this stage be willing to start 'badly' because you have enough time to improve. Even if you're talking at short notice, you have to start somewhere – just like everyone else does. Make the bar so low for yourself that you'll hop over it with no problem, then everything after that can only get better.

Rehearsal

Once you've practised on your own, try if at all possible to rehearse in as similar a scenario as you can to the actual event. You want to make it as real as possible so it's not a shock to your system on the day. Can you use a large auditorium late in the evening when it's empty? Can you borrow a space from your university? If you're giving a TV interview, can you practise on Skype or Zoom and ask someone to pose you questions? Put several timeslots in your diary for your rehearsals and don't break them – they're an investment in yourself that will pay dividends.

Your voice

When you're rehearsing, pay attention to how you deliver your content and to the tone of your voice. So often I see presenters talk in a formulaic, repetitive and monotonous tone and rhythm, which isn't likely to be memorable and certainly isn't giving any evidence of energy. Listen to yourself and be honest if you hear yourself falling into this trap.

To avoid it, think about your presentation as though it was broken up into short sections with each part having a different tone. Your greeting is about energy and connection ('I'm so excited to unveil this innovative research approach today'); the introduction to your main content is about curiosity and clarity ('Let's look at the study design – there are three things that are important here'); the results are about discovery and insight; and the ending is about confidence ('I feel confident that this research will transform the lives of thousands of people. Thanks for listening.'). You could think of your talk as having chapters rather than slides, with each one having its own tone of voice.

Consider using pauses and repetition to emphasise your key messages: 'We saw that there were twice as many patients who benefited from this medicine. [pause] Twice as many!' Remember that your audience members are hearing this for the first time, so they need that pause and repetition to appreciate the importance of what you're saying.

Here's a checklist of what to watch out for in your vocal delivery:

o Does your voice sound rich and relaxed, or tight and stressed?
o Have you warmed up your voice enough?
o Notice your pitch. Does it sound authoritative? Do you want it to be higher or lower?
o Are you speaking too quickly or slowly?
o Where do you want to slow down and where might you speed up?
o Does your speech sound energetic and lively, or dull and monotone?
o Is your volume enough? And is it the same throughout, or do you vary how loudly or softly you're speaking?
o Are you emphasising the key messages?
o Are you using pauses effectively?
o Are you speaking in short phrases that your listeners can easily understand?

o   Listen for where you pause for breath, because we speak on the out-breath. If you run out of breath, slow down and break down your utterances into smaller chunks.

Rehearsing helps with both your vocal and physical preparation as it also has the benefit of reinforcing your vocal muscle memory so you remember your words, strengthening your brain-to-voice connection. I call it 'mouth-feel'; the more you sense the phrases orally, the better you deliver them and the tighter your communication becomes. That's how good speakers come across as natural; they're comfortable with their words, so they're comfortable with speaking them in any situation.

Your body language

When reviewing your video recordings, also check out your body language because the way you present yourself physically is important. Remember that your body language needs to be congruent with both your voice and your words in order to build trust and create an impact. Take a look at how you're using your eyes, hands and posture. If you're presenting on stage, are you facing your audience or are you staring back at your slides or down at your notes? Looking at the audience is essential for connecting with them.

How are you standing? Your stance can convey either nerves or confidence. Even if you're nervous, you can stand in a way that makes you feel and appear more confident. Ideally you want to be upright and standing, or sitting tall with your shoulders relaxed. If you're standing, place your feet far enough apart to give you stability; this is usually a little more than hip distance apart. If your feet are close together, you'll have a small centre of gravity and are likely to rock from side to side or from back to front. This is something you might not notice until you watch yourself on film, but it can be distracting for your audience.

How you move will often depend on the type of talk you're giving – for instance, wide arm movements can work well on stage if used appropriately but would appear too expansive on camera. I go into more detail on this in the next chapter, but in the meantime here's a checklist for your physical delivery:

o   Are you standing or sitting up tall?
o   Are your shoulders relaxed and down, or do they appear to be hunched forward?
o   Do you look comfortable and relaxed in your space?

- Are you standing with your feet wide enough apart to give you a solid base, or are you rocking side to side or backwards and forwards?
- Are you facing your audience with an open body?

Finally, if you're still wondering whether all the hours spent practising and rehearsing are worth it, it's helpful to know that learning changes your anatomy. Our brains contain neural connectors called axons, which are long, threadlike parts of the nerve cell that conduct impulses to other cells. As they're stimulated, they grow thicker myelin layers, making the pathways stronger. Practice and rehearsal strengthen these pathways so they become the default route for sending neural signals to the rest of your body (this is why sports players and actors constantly rehearse specific skills). When you rehearse, you physically grow your expertise at speaking. And yes, like anything worthwhile, it takes work and a certain amount of courage because beginnings are usually awkward. But trust me, it's definitely worth it!

Physical and vocal exercises

Because speaking is a physical activity, it demands physical preparation. This is why exercises are invaluable for helping you to relax, and to prepare for the energy required to deliver a compelling performance. They're also a soothing and empowering ritual to enact before a talk, because they bring you into the present moment and make you focus on yourself. Ideally you'd practise them for a few weeks in the run-up to your big day, so that when it arrives you can run through an edited selection and they'll feel like second nature. At the end of the book, you'll find a special chapter ('Physical and Vocal Exercises for Speaking Success') that walks you through my suggested exercises one by one, so rather than go through them here I suggest you turn to that chapter when you're ready and familiarise yourself with them. You can also see videos of the exercises at https://scientificallyspeaking.com/physical-and-vocal-exercises.

One size doesn't fit all. Some people find breathing exercises helpful, while others prefer body-based ones. I'm a fan of power poses, and can often be found in the Ladies before a talk with my hands in the air like Usain Bolt or Wonder Woman (or in my living room if it's an important webinar I'm about to host). Experiment with what works for you, pushing yourself a little beyond your comfort zone; once you're comfortable with one exercise, move on to the next.

Mental preparation

When you're nervous, it's natural to think in a negative way. You might be saying to yourself, 'I don't want to be here, I wish I could walk away. Everyone will be staring at me, criticising my every word. I'll forget what I was going to say. I'll make a fool of myself. It'll be awful.' Written down like this, it seems a bit ridiculous, but it's common to think like that.

One of the best ways to counteract this negative spiral is to replace your anxious thoughts with positive ones: 'I'm here now. It's a given that I'll walk onto that stage in an hour. My aim is to give it my best shot, and even to enjoy it.' Remember that this is your opportunity to shine, so step into the spotlight and own it. When you tell yourself you *want* to be there, your body language starts to reflect that and you'll come across as confident and in control.

What's more, let's entertain the prospect that the experience might not be akin to a torture chamber. You might be envisaging a sea of audience members frowning with their arms folded, so let's flip the image. Be deliberately optimistic. Imagine these people are already your friends, excited and inspired by what you've just shared with them. Everyone will have a smile on their face because they're interested in what you have to say. And start to envisage not only surviving the experience but even enjoying it. This way you will reframe your ambition so that your purpose isn't just to get through the talk alive or not to say something stupid, but to influence people and engage with them.

On the day

What happens on the big day itself? You'll be pleased to know that there are practical steps you can take to make your presentation, video recording or interview go well. Not only are they beneficial in their own right, but they'll also help you to feel calm because they will give you something constructive to do.

Check your tech

Make sure you check your technical equipment – do it well in advance. It's hard to display energy if your laptop battery runs out part-way through your presentation, or you can't find your clicker. I once ran an intensive, three-day training session for more than 30 doctors. We were a team of four trainers, and as we set up on the first morning we discovered to our horror that none of our six computers would connect to the projector, which meant that we had no way of showing our slides and videos. After an hour of frantic problem-solving we managed to crack it two minutes before we were due to start. I was the first person on, and anyone who knows me would have noticed that I was less polished than usual because I hadn't had a chance to do my calming and energising exercises beforehand.

What if you're not responsible for the technology yourself? Try to insert yourself into the event's checking processes. For a virtual presentation or panel, arrange for a technical run-through in advance and make sure your video and audio set-up works. At a physical venue, go to the slide rehearsals and survey your slides, making sure they're in the right order and that they show up well on the screen; test the microphone and check that the lectern (if there is one) is in the right place for you.

Prep the scene

If you're at a congress, scope out the room so it's familiar, and see whether you can find a time to stand on the stage to experience what it feels like when the lights are on you. The same goes for a TV interview – find out where it is and even (if you can) ask the interviewer what your first question will be. Discover as much as you can in advance because the unknown is always more frightening than the reality.

Have water to hand

Do you get a dry mouth when you're nervous? If so, have some water beside you. I always have two litre bottles of water on my desk when I host a webinar, because I know that a dry mouth when I speak will give away my nerves.

Pep yourself up

You can do some of the physical and vocal exercises at the end of the book, most of which are suitable for any venue as they're either invisible or can be done in the privacy of a bathroom. These will help you to feel relaxed and energised. Some people also like listening to their cheesiest and most uplifting songs on their headphones.

'Friendify' your audience

When people concentrate, they often frown and look serious; this is why, when you speak to a large audience, you're often presented with a sea of what I call 'scary learning faces'. However, they're not necessarily bored or annoyed, they're just focusing hard on you. Something else that can be off-putting is when you see people on their phones. Of course, it would be lovely if everyone was polite enough not to do that, but you have to accept that it will happen (and they may be tweeting about your latest point). When speaking to an auditorium, I suggest asking two or three of your friendly colleagues to be in the audience to support you – one on the left, one on the right and one in the middle – and asking them to smile at you and look interested the whole way through. Then anchor your talk to them. Or if you find all the faces too much, fix your gaze about 30 centimetres above people's heads and speak to the back (you can have an imaginary best friend situated at that point).

Protect your time

The hour before you speak is a precious opportunity to focus on you and your talk alone – I call it sacred time. Don't look at your phone, discuss anything worrying or contentious with anyone, or start talking to yourself in a negative way. Instead, do some exercises and surround yourself with people who make you feel good. Alternatively, if you're an introvert and find company draining, limit your contact with others so you don't have anyone fussing over you. Think about what you honestly need and be selfish. If you require energy, do some voice-warming exercises to make sure your voice isn't small when you start speaking. And if you feel extremely nervous, do some slow, deep breathing or other conscious physical relaxation. See the 'Physical and Vocal Exercises for Speaking Success' chapter at the end of the book, and online resources that will help you with this at https://scientificallyspeaking.com/physical-and-vocal-exercises.

You can see how all the work you've done ahead of time to practise and rehearse what you're going to say comes into its own now. Instead of worrying about how you're going to introduce yourself or explain a complex theory, you can look directly at people, smile and engage with them through your positive body language. They'll smile back at you in turn, and you'll feel better. The more present you are with them, the more your one-way talk will feel like a two-way dialogue because they're *with* you, not just in front of you.

This is what speaking is all about.

The main points

o   Practising and rehearsing are the two most important elements of preparation because they enable you to give a polished performance with energy and confidence.
o   Speaking is a physical activity, so you need to exercise your body and voice beforehand.
o   Preparing your mind to feel positive and upbeat will give you energy on the day.
o   There are various things you can do on the day itself that will help you to feel calm and in control.

Chapter 9
Speaking in different settings: Virtual meetings and other formats

Television is for appearing on, not for looking at.

Noel Coward

Throughout this book, I've mentioned the different scenarios in which you might be asked to speak: on stage, in a media interview, on a panel, and online (to name a few). Together these make up the 'In which channel?' part of the communications model that we explored at the start of the book (Figure 10), and each of the different channels has its own specific characteristics.

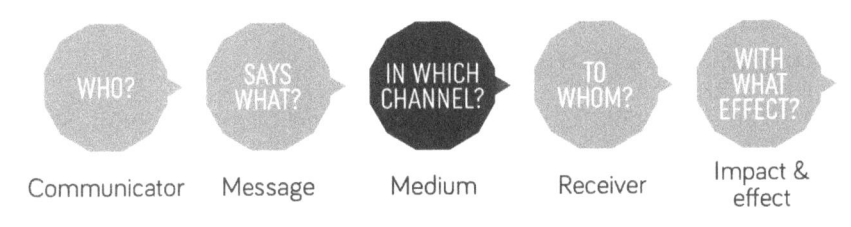

WHO?	SAYS WHAT?	IN WHICH CHANNEL?	TO WHOM?	WITH WHAT EFFECT?
Communicator	Message	Medium	Receiver	Impact & effect

Figure 10: It's important to understand the characteristics of each channel

Because of this, they also require different skills: presenting on stage is a different experience from giving a webinar, and answering questions on an expert panel needs to be handled differently from responding to questions from a journalist. Just as you need to tailor your speaking to the audience you're addressing, you also need to adapt yourself to the channel in which you're speaking.

Of course, the COVID-19 pandemic caused so much to change regarding speaking situations. Online video conferencing has now become the norm and in-person symposiums are rare; now that so many of us have experienced the flexibility and convenience of speaking online, it's hard to envisage us returning exactly to the way it was before. For this reason, I've devoted much of this chapter to communicating online – there are many new skills to learn in this area.

After that, I'll also cover what you need to know in order to speak to camera – for instance, if you're recording a message to colleagues or a video to post on Twitter, giving a media interview, being interviewed for a podcast or presenting on stage. In doing so, I'll be drawing on my personal experience as an interviewer, spokesperson and interviewee, as well as my years spent training countless doctors and scientists. I approach the topic in an intensely practical way, so by the end of this chapter you'll have all the information you need to feel confident about presenting in pretty much any scenario.

Before we begin, it's important that you've already read the 'Mastering the Q&A' and 'Managing Your Nerves and Speaking With Confidence' chapters. So much of what you learnt there is relevant for this one, because understanding how to answer questions and getting yourself in the right physical and mental state for speaking are both essential for an impactful delivery.

Video conferencing

Many of us have spent more hours than we can count on Zoom, Skype, Microsoft Teams, Google Meet, or some other kind of video conferencing platform over recent times. As I'm sure you've realised, there are some people who come across well on these platforms and others who appear ill at ease, hard to see or difficult to understand. Yet today there's more of a need than ever to create connection and impact in the virtual space. This can be problematic when you consider the short space of time we've all had to adapt to these new platforms, so let's look at what you need to know if you're to come across confidently and persuasively online.

First, I should be clear that all virtual meetings are not the same. In fact, I identify three types:

1. *simple* (such as a catch-up with your colleagues or team – a meeting that's interchangeable with a phone call or face-to-face meeting)
2. *complex* (such as a job interview, advisory board, brainstorm, training session, or a meeting with many people and possibly a long agenda), and
3. *broadcast* (a large scale, one-way communication such as a conference presentation).

To keep things simple, I'm focusing purely on the broadcast here (think of a medical congress presentation, for instance). This usually consists of either live or pre-recorded slide presentations; in the case of pre-recorded presentations, the talks are often videoed in advance, with the Q&A being broadcast live on the day of transmission, sometimes as a panel discussion. Much of what you'll learn for broadcasts will also be useful for the other types of video meeting, because there are some key considerations that are distinct from speaking in person.

Winning the war of attention

It is important to bear in mind that everything in virtual presentations is about winning the war of attention, a battle that is even more acute online than it is in person. Remember what it was like when you last sat in front of your screen watching a live talk? You probably had email or update notifications popping up, your phone might have started buzzing or flashing with messages, you could have had another screen on your desk such as an iPad, and you probably experienced interruptions such as the doorbell ringing or your children coming in. If you're the speaker, your audience is no different.

It's the same when you're watching a video online. What's the first thing you do when you click 'play'? If you're anything like me, you look at the bottom right-hand corner to see how long it is. If it's too long, you may decide not to watch it at all, and even if it's a short video and the presenter doesn't grab your attention within the first few seconds, you click away. This is no exaggeration; research shows that most people turn off a video if it doesn't hook them within the first eight seconds, and the remainder by 30 seconds.

This means that it's essential to launch into your topic in a compelling way. If you start with, 'Hello, I'm in sunny Barcelona at this amazing conference,' you've lost them at hello. Your audience wants you to be brief and to the point.

Everything I'm about to tell you should be understood in the context of focusing people's eyes and ears on you and your content. With that in mind, let's look at the various elements of presenting online that can cause you problems, and how to solve them.

The technology

Using technology to present is nothing new, in that when you speak on stage you have a microphone and a slide projector to work with as a minimum. However, there's a technical team to set it up for you. With video conferencing, you're usually managing it yourself from your home office, which means that many of the frustrations you can experience revolve around difficulties with using the tech. The result of this can be a poor impression on your audience as they watch you squinting at the screen while you wonder what button to click, or struggle to hear what you're saying as your broadband dips in and out. Given that it's just as important to be fully present with your audience online as it is in person, you need to feel confident about the tech. If you're not, you'll feel stressed, your attention will be divided and the people you're addressing will be less likely to engage with you.

First, for important presentations use the best wi-fi available to you and, if you can, hard-line it into the device you're presenting on. Make sure you've downloaded any updates for the video conferencing software and be familiar with its functions. If someone else is hosting the event and you need to be able to share your screen, ask them for permission to do that from their end in advance. You'll diminish your initial impression if the beginning of the meeting is taken up with sorting out this kind of technical issue.

When it comes to your content, rehearsing with the technology is just as important as practising for an in-person event. Just as you shouldn't be saying the words for the first time on the stage, so you shouldn't be using the tech platform for the first time either. When I give online presentations that are especially important to me, I plan at least three rehearsals with different groups of people on the specific platform I'll be using on the day. By the end of the rehearsals, I know where I need to click and when, which slide comes next and how I plan to interact on the chat. I feel confident, which means I'm fully present for my audience – and it makes a world of difference.

Gaining attention and reducing distraction

When I give video conferencing training sessions, I start with a piece of footage that always makes people smile. It consists of a news announcer who's sitting in front of his mantelpiece. So far so professional, but what he hasn't noticed is that his ginger tabby cat is curled up there having a snooze. As he reads the news the cat slowly unfurls and starts licking its bottom, an activity it continues to enjoy until the end of the piece. Whenever I ask people what they notice about the film, they always say the cat – usually with a euphemism about its activities. But when I then ask them what the newsreader said, no one can tell me; this shows how impossible it is for your audience to listen to you if there's something visually distracting going on at the same time.

This is key to remember, as it applies to all the advice that follows. There are many ways of participating in a video presentation. There are also many places to position your camera and techniques for adjusting your lighting. If a meeting is low stakes, you might not need to worry about how you're coming across; however, if it's important, you'll want the attention to be firmly on you and your message. So when I talk about your background and lighting, it's not because there's only one right or wrong way to do it, but because there are things you can do with these variables that attract attention to you, and there are things you can do with them that detract. How you stage and set up your video equipment makes a crucial difference to winning the all-important war of attention.

Background

The background to your video is the stage you set for yourself, and it sends a message in its own right. Most people want this to look either neutral or professional, as if it's a study (even if it's not); think of it as if you were a newsreader broadcasting from home. Ideally you'd start with a blank wall and then add a plant, a tidy bookcase, or picture to make it appear more furnished. The aim is for people to look at it for a couple of seconds, think it's appropriate for you to be speaking in and then ignore it – not to be wondering what's behind the semi-open door or whether that's an ironing board in the corner. Kitchens are especially distracting because people love to have a nose around: 'I wonder where they bought those tiles' or, 'Ooh, I've always wanted one of those coffee machines!' The good news, of course, is that you can prepare your background ahead of time – in fact, you have far more control over your environment than you do on stage, so take some time to make it work for you.

Lighting and sound

LIGHT FROM BEHIND THE CAMERA
Natural light or a desk lamp behind the camera allows people to see your face well

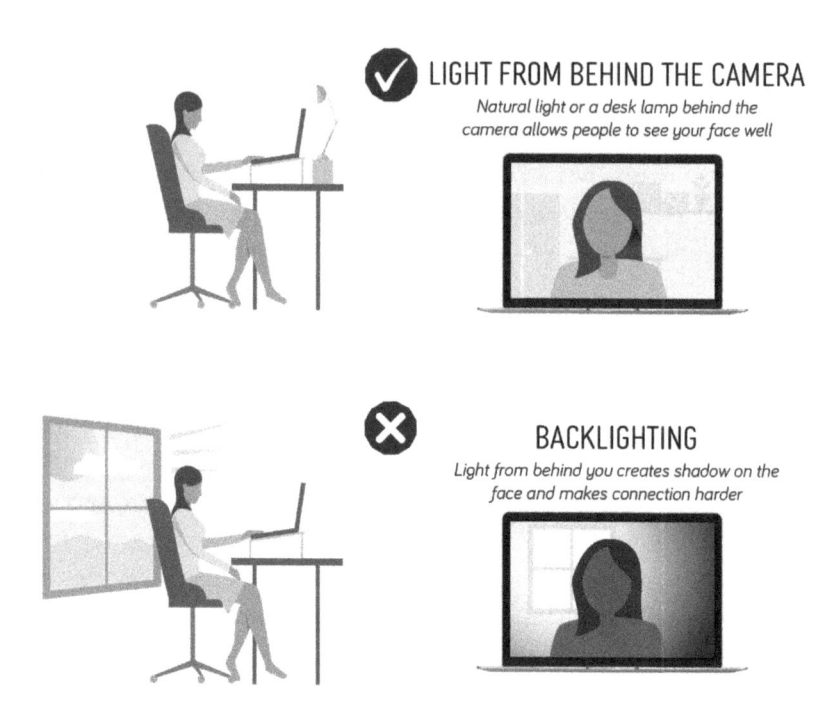

BACKLIGHTING
Light from behind you creates shadow on the face and makes connection harder

Figure 11: It's important to ensure that you light your face well from behind the camera

If people are to feel connected with you, they have to be able to see you, so you need good lighting on your face. Much of the story you tell with your talk comes from your facial expressions, and if you're hidden in shadows you'll lose a lot of impact. The main rule is to have light shining directly on your face, not from the side and definitely not from behind (a window behind you will create a grey silhouette) (Figure 11). Natural light is ideal, but if your window is in the wrong place or you know that the light will change during the course of your talk (such as during an evening presentation), lamps are helpful. I use two selfie ring lamps that clamp to my desk, one on each side, and they ensure that I'm always presented in an even and well-lit way.

With sound, your aim again is to avoid creating unnecessary distractions for your audience. If you think background noise from your home or office will be an issue, you can use a headset (this will have the added bonus of helping you to concentrate). A headset can also help to avoid echo in a large room.

Camera height

Visual framing is important. The standard way of presenting, as I'm sure you've seen when you've watched talks by others, is to put your laptop on your desk and talk into the camera. However, the lens will almost certainly be too low. Your viewers will have an unflattering view up your nose, they'll see a lot of ceiling and you'll appear as if you're looking down on people, which can seem patronising. Also, if you're dropping your chin you're restricting your voice box, which weakens the impact of your delivery. At the other extreme, some people put their camera up high so they're tipping their head back as they look into it (this often happens if they have a camera clipped to the top of a large monitor). This technique comes from the belief that taking a selfie is more flattering when you're looking up, but in a conference call it just looks weird. You'll also see a lot of floor that way, which is distracting.

The ideal position for your camera is at eye level. You want to frame the shot so that your eyes are about one third down from the top of the screen, so there's not a huge gap between the top of your head and the top of the frame and your face is straight on to the camera (Figure 12). This will almost certainly require you to raise up your laptop on a stand, a stack of books or a box. My husband uses an upturned laundry basket – it doesn't matter what you use as no one will see it. Take time to experiment with different heights, as it makes a huge difference to how professional you look.

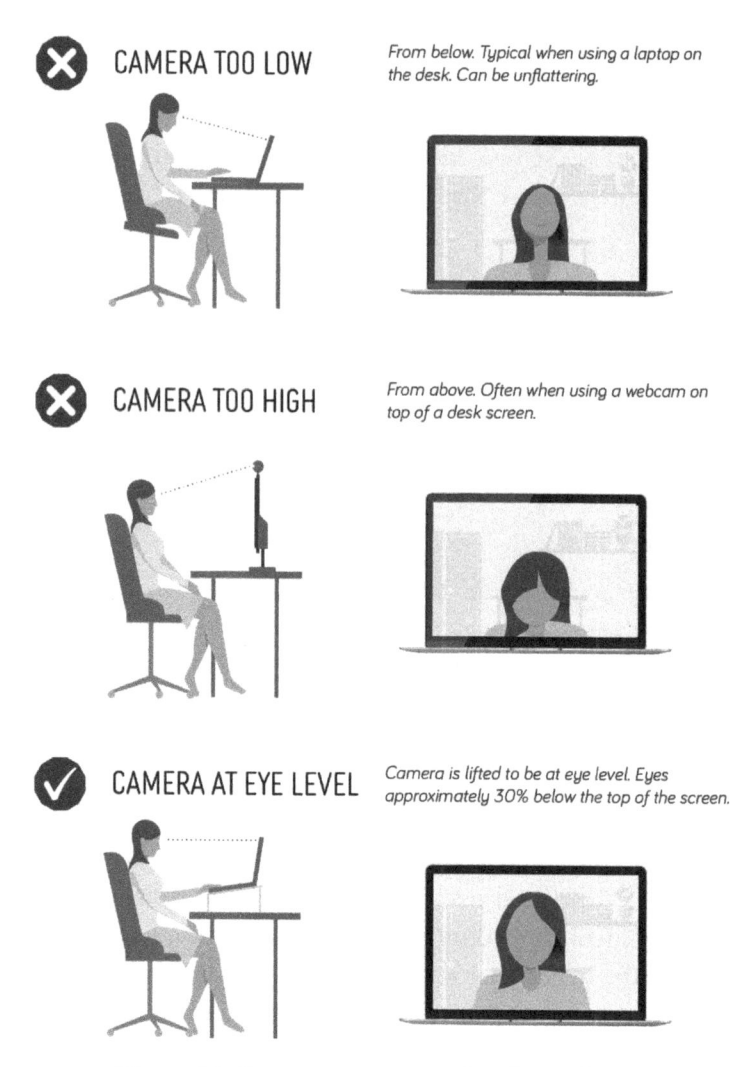

❌ **CAMERA TOO LOW** — From below. Typical when using a laptop on the desk. Can be unflattering.

❌ **CAMERA TOO HIGH** — From above. Often when using a webcam on top of a desk screen.

✓ **CAMERA AT EYE LEVEL** — Camera is lifted to be at eye level. Eyes approximately 30% below the top of the screen.

Figure 12: Ensure your camera level is in line with your eye line

One final tip about camera set-up for Zoom or Skype television interviews: if you're being interviewed remotely through a conferencing system, ask the producers whether your screen will be shown in portrait or landscape mode. Landscape is most common, but if you'll be displayed side by side with someone else it may be portrait and you'd be better off using your phone than your computer. There are online resources available that can help you with this in more detail.

Eye contact

Just because you're sitting in your home office talking to a screen, it doesn't mean that people on the other end aren't looking for you to make a physical connection with them. The main way you do this is through your eye line, by looking directly at the camera. In fact, you want to look 'through' the camera, as if your audience was sitting directly behind it – you can stick a piece of paper with a smiley face in front of your camera, with a hole-punch hole for the nose and lens, if you like (Figure 13).

Figure 13: A reminder to maintain your eye contact with the camera

Because it's human nature for your eyes to be drawn to the faces on your screen, or to your slides, you have to train yourself to look at your camera instead. On a conference call, you can make it easier for yourself by minimising your screen and dragging it as close as possible to the lens, so your eyes don't have to travel far from one to the other. If you're presenting a slide deck, this won't be possible as you'll need to see your material, but you can still look at the lens as much as you can.

Another way of bringing eye contact into your talk when you have a live audience is to stop screen-sharing every now and then to take questions, or even just to explain a point in more detail. You don't have to share it the whole time (I insert a blank slide into my presentations to remind me when to pause sharing).

Your body language and energy

Speaking online is a different experience from speaking in person, but it's still important for there to be congruence between your content, tone of voice and body language. In fact, you need to talk with more energy than if you were speaking to someone in the same room, as you have to bridge the technology divide. Imagine addressing a large meeting and you've probably got it about right. You're not shouting, but you're putting energy into your voice. That's why it's important to do some physical and vocal warm-up exercises beforehand, especially if you've been at home all day and haven't used your voice before the meeting (refer to the 'Physical and Vocal Exercises for Speaking Success' chapter at the end of the book).

Your delivery should demand people's attention even more than it does on stage because of the extra distractions with which your audience will be dealing. You have to be passionate about what you're presenting, even if you've said it a hundred times already. Connect with your purpose, smile and deliver with enthusiasm. In fact, smiling makes a huge difference to how people relate to you. And if you're not speaking for a while – for instance, if you're on an online panel – think about how your face looks. I've previously been given feedback that I sometimes have an expression that appears either bored or angry when it's in resting mode, so work hard to look engaged.

Some people wonder what to do with their hands when presenting virtually, but you can use hand gestures in the same way as you normally would; they will give energy to your performance and increase your impact. If you tend to use your hands a lot, position your camera far enough away that your hands can come into your frame without looking as if they're flying into the screen from nowhere. Make sure you know the size of your frame so you can work within it. If you don't envisage using your hands that much, keep them out of shot. Should you want to bring them in, you can do so and then put them down again.

If you're on a video panel and you want to interject to make a point following on from another speaker, you can use your body language to signal to the moderator that you want to talk. You might lean forward or raise your hand. When you come in, do it assertively: 'I might add something there' or 'John, I'd love to add something.' Think of the tone of voice you'd use if you were greeting someone and begin speaking with confidence.

Finally, remember that if it's an important presentation you'll probably feel nervous and that nerves can show up as a lack of authority. So practise your talk until you know your content forwards and backwards, and do some physical exercises beforehand. It can be challenging to generate the energy you need when you're sitting at home and staring at a computer camera, so you have to find ways of pepping yourself up as if you were on stage. This is your moment, so step into it and own it.

Standing or sitting?

A lot of people ask me if they should stand up when giving a piece to camera. I always say that the most important thing is for you to feel comfortable. Again, think about the impression you want to create and how you'll be framed. If you're sitting, you'll want to show yourself from your shoulders or maybe even from your waist. If you're standing, it can be the same, but with the option of showing your whole body if you want (and your camera will need to be placed at a high level; you might need to put it on a chest of drawers or other tall item of furniture).

You might prefer to sit for self-recorded video content as it helps keep you in one place and framed as you planned. Sit up straight from the waist and avoid leaning back on your chair because this looks too casual. Also avoid creating distraction by swivelling on your seat.

Some people prefer to stand to deliver an online presentation, and if your talk is no more than about 10 minutes long, this can work well. Any longer and you may start to feel tired, which will reduce your energy. The key challenge with standing is the temptation most people have to shift their balance from one foot to another in a distracting way, which won't help to win the war of attention in the right way. If you choose to stand, do make sure you have a wide stance to avoid the chance of rocking.

Managing interruptions

The ultimate distraction, both for you and your audience, is when one of your children enters the room, the doorbell rings or some other unpredictable event occurs that knocks you off your stride. Of course, you'll no doubt do your best to ensure that everyone knows you're not to be interrupted, and that there's someone to make sure the dog doesn't bark, but sometimes life happens and you have to deal with it professionally.

The best thing to do is to embrace it; don't pretend it's not happening. 'I see we have a guest!' you might say if your child enters unannounced. You can then choose how to manage the situation, either by welcoming the interruption into the session (because that can be less distracting in the long run) or by saying something like, 'Excuse me for a moment while I deal with this', then turning off your video and audio while you do so. The main thing is to have a plan and a few phrases ready to use, because if you're relaxed about interruptions everyone else will be too. Most people understand the challenges of working from home and are sympathetic. An interruption from a child or pet, if managed with comfort and confidence, can actually work in your favour if it helps to build connection with your audience. Even if it's not professional in the conventional sense, it offers you a chance to show your humanity and sense of humour.

You can download a tip sheet about presenting online at https://scientificallyspeaking.com/virtual-presenting-tips.

Recording to camera

There are various reasons why you might want to record a short talk to camera: for social media, a training video, or some information for your colleagues. You'll find that in the online world of social media, you gain a lot more attention and engagement if you share a video than a written post, so if you've not done this before it's well worth considering. When it comes to making an impact, much of what you've learned with video conferencing, such as having good light and sound, is all equally relevant. However, there are also other considerations, which I'll cover here.

Body position

Many videos recorded to camera with a professional support crew are carried out with the presenter standing up, as it gives a spontaneity and energy that's reminiscent of a stage presentation. If you're self-recording, however, you can choose to do it in the way that feels most comfortable, and this is often sitting down in front of your computer.

Again, the same principles apply. To avoid distracting people, avoid moving around a lot. If you're standing, place your feet far enough apart for it to feel unnatural. This is to give you a secure base, which will stop you shifting around because it requires a significant effort to move from this position. When you have your feet close together, your centre of gravity is small and it's easy to move from one side to another without realising it (or even to rock back and forth; place a wedge under one of your heels if you're prone to this). Obviously, if you're filming your whole body you won't want to stand with your feet too wide apart, but it's helpful if you're shooting from the waist or chest up.

If you're standing and filming, it's a good rule not to drop your hands by your sides, where they can look limp, or to grasp them too tightly in front of you, which puts tension into your shoulders. Instead, keep your hands with a little energy in them resting at belt level in a neutral position, with your fingers together. If you naturally 'talk' with your hands, you can do this on camera – just be aware that gestures that are too expansive can also be distracting. It's helpful to record yourself and practise so you're comfortable with how you come across.

Deliver with energy

So much of the impact you make when speaking to camera comes from your delivery. Look 'through' the camera, imagining that there's someone you like and trust on the other side, and present to them with energy. If you find this difficult, see whether you can ask someone to stand behind the camera for you and speak to them through the lens.

If it's a short piece that you're recording, you don't have the luxury of warming up as you get going. So before you press record, do some vocal and physical exercises to put you in the right frame of mind and ensure that your voice is ready to go. Then smile and begin with enthusiasm, carrying on the momentum for each new section of your talk. As with in-person speaking, it's good to vary the tone so when you shift to a new thought or topic your audience can tell that you're addressing something different. This will help them to stay engaged.

If you're feeling a bit nervous, you may find that you speed up your delivery without realising, so make a conscious decision to slow down. This gives people time to absorb what you're saying. Using short, clear statements also makes your messages easy to understand; don't think that you're giving a lecture – you're focusing only on one short section at a time and telling each part of the story before moving on to the next.

Know who you're talking to

Are you addressing people who will be watching together in a group – for instance, in a video message to people at a meeting or congress that you can't attend? Or will members of your audience be watching you individually on their own? Video messages are both public and intimate, so you want to create the impression that you're speaking directly to a colleague or acquaintance. Most of the time we watch videos alone on our computer screens, so 'Hello everyone' doesn't usually work. In fact, it can be a little distancing. Only do the 'Facebook Live' thing of addressing lots of people at once if you're sure you're talking to a group, or if you are indeed doing a Facebook or Instagram Live and are interacting with your audience in real time.

Clothing choices

Here it's important to think again about the war of attention. You want people to focus on you and your message, not on your outfit, so something smart and simple is often best. Small and closely mixed patterns, whether they be on your tie or your top or dress, can create a 'buzz' on screen and are best avoided. Also be aware that, for men, solid-coloured shirts generally look better than white. For women, structured shirts, jackets or dresses tend to look more professional on screen than cardigans or sweaters.

Media interviews

As with all the skills we've discussed, preparation is key to a successful interview. Most of what you've learned about handling the Q&A also applies to managing media interviews. Think about which questions are most likely to be asked and develop a plan for how you'll respond to them.

It's worth knowing that even if a journalist calls you and asks for an immediate interview, you don't have to do it right away. It's okay to ask for time to prepare. Try to negotiate as much time as possible, even if it's urgent – you can do a lot in 15 minutes. Identify your goal, write down and practise speaking your key messages, and check your facts and dates. If you're asked to comment on new data with which you're unfamiliar, ensure you've had time to review it so you can form an opinion and articulate your message in advance of the interview.

Remember you're in charge of what you say – a journalist can't put words in your mouth. If they ask a negatively framed question, you don't have to answer it using the same vocabulary – it's not an exam. Instead, respond with what your audience needs to know. Keep your answers brief and in a conversational style, avoiding jargon.

If you have the opportunity, it helps if you can find out as much as possible about the interview beforehand. Are you the right expert for the interviewer (you can always decline an interview if not, and perhaps suggest a colleague or another expert), and are you happy to talk about the topic they have in mind? Do you have a strong message to impart that's relevant to it? Assuming all is well, see whether you can find a way to bring in that message as early as possible because you don't know how long the interview will last. If your aim is to convince your audience to get a particular vaccination, for instance, don't wait until you're asked about it. If the question is, 'So how safe is this new vaccine?', you can answer, 'This vaccine prevents the spread of a dangerous disease, so it's vital that people make an appointment to have it.' Remember that it's not really the interviewer you're talking to but the people who are watching or listening at the other end.

Also, don't be afraid to repeat a previous answer using different words. This is where your stories, sound bites and statistics can come in, as they will make the repetition interesting and memorable.

Your location

There are a couple of considerations if you're being filmed in locations that aren't on stage or a studio. If you're outside be mindful of what's behind you, and if you're at a congress on behalf of yourself rather than a company or institution, make sure you're not standing in front of a corporate stand, as this could give the wrong impression. You have agency regarding this, and it's important that you're happy with how you're being framed, so don't be afraid to check this with your interviewer.

Podcast interviews

Being interviewed on a podcast is an excellent way to build your influence online and to spread your message widely. This is an intimate medium, so the quality of the sound is important. Try to sound as clear and energised as possible, as when people are listening to your voice without the benefit of being able to see you, they'll draw conclusions about your authority from your tone alone.

Before the interview

You're the expert and you're sharing important views, so you need to sound authoritative. Carry out some warm-up exercises so you're feeling positive and calm. Also make sure you have excellent sound quality; this might involve investing in a high-quality microphone or a set of headphones. You don't have to spend a lot, but speaking from an echoey room with intermittent background noise will be highly distracting for your audience.

During the interview

A podcast interview is usually more conversational in tone than a news or journalist interview. The medium lends itself to longer conversations and a relaxed feel, but that doesn't mean you shouldn't prepare just as thoroughly as you would for a media interview. Given that you'll be talking for longer, storytelling, anecdotes and analogies are important, as you want to engage the audience and bring them into the world of your expertise.

The good news about podcasts is that because you're not usually visible (although some are videoed), there are liberties you can take that aren't possible in a visual recording. You can print out your notes and have them to hand, although don't put them on your desk because you'll be looking down at them and this weakens your voice. When I was a spokesperson giving telephone interviews (before the days of podcasts), I used to stick my notes to my window at eye level so I could walk up and look at them. If it was a long conversation, I might even check off what I'd talked about so I could see what was left to say.

It's up to you, but it's perfectly okay to stand up throughout if this helps you to feel confident and at ease (you'll need a headset if you do this, otherwise you'll move in and out of microphone range). You want to sound passionate and to have energy in your voice, so feel free to use your hands as you normally would because this will lift how you come across.

However, having energy doesn't mean speeding up. Slowing down is even more important on a podcast than when you're recording visually, especially when you consider that some of your audience may not have English as their first language. Everyone needs time to absorb your messages, and it also gives you time to think and makes you sound authoritative. If you tend to gabble when you're nervous, try putting a note on your screen reminding you to 'slow down'.

Speaking in person

If you're speaking to an auditorium from the stage, you're engaging in live theatre. The content of what you say matters, but the main source of your impact comes from your physical presence and your voice. You need to engage with your audience from the moment you step onto the stage until the moment you leave, and this takes preparation, practice and rehearsal – much of which we've already covered in detail. What we haven't been through, though, are some of the nitty-gritty elements of live speaking with which you may not be familiar.

Body language

First, when you walk onto the stage, do it with confidence and acknowledge the audience in some way if you can. This first moment is one of the most nerve-wracking, so you might need to remind yourself to smile and look at people, focusing on the most friendly faces. During your talk, you'll be tempted to keep looking at your slides on the screen behind you, but you'll almost certainly be able to see them on a monitor in front. Don't turn around to point at them unless you absolutely have to – it's far better to maintain eye contact with your audience. And don't read the slides aloud because people can do that for themselves more quickly than you can speak them; instead, tell them the 'so what' and bring your perspective and insight into play.

Also, remember that when you're one person on a big stage, your body language should be more expansive than it would be in any other situation. You can take up a lot of room by widening your arms and making large gestures.

Tech considerations

If you're in a large auditorium, you'll probably be given a lapel or over-ear microphone to wear. If you're a woman, it's worth bearing in mind that mics were designed for men. So it's a good idea to wear trousers or a skirt rather than a dress, or if you do wear a dress have a belt for the battery pack to fix onto, otherwise you'll have it hanging off the back of your neckline, which is uncomfortable. Similarly, a jacket is a more secure location for a lapel mic than a cardigan or unstructured piece of clothing (and floaty scarves can interfere with the sound quality).

In addition to microphones, you might also have to take into account the presence of video monitors, which show your face up close for people who aren't at the front, or for online sharing purposes. In fact, most people will see you this way. It's a good idea to scope out the room first so you can see where the cameras are, as this will help you to look at them from time to time. Not only will you then be connecting with the majority of your audience, but it's a lot easier to look into a camera than a sea of 5000 faces.

What to wear

You need to look the part, so dress in the way that's appropriate to how you want to come across. And make sure your clothing is comfortable to wear. I usually tell people to choose an outfit in which they feel great. You want to project with energy, which is hard if you're constantly tugging at your waistband or smoothing down your top. You could ask a friend to take a picture of you ahead of time so you can assess your outfit, which is especially important if you're going to be sitting down in it (see below).

Don't be distracting with your clothing choice by wearing anything too fussy or detailed. If you're a woman, you have the advantage of being able to stand out with your outfit, which helps you to be remembered – that's why I always encourage women to wear a colour rather than grey, or black and white. Another tip for women is to wear trousers rather than a dress or skirt, especially if you're sitting on stage in a panel or during an interview. Some panels have high stools or low armchairs, which can cause your skirt to rise up, and I've had to go on emergency clothes shopping trips for clients in the past to avoid any wardrobe mishaps.

You might be finding the range of these considerations a bit overwhelming, but they're not designed to be learned by heart. Instead, just refer to the appropriate sections as and when you need them. The key points to bear in mind are that effective speaking is a mixture of the right content and the right delivery, and that your techniques should vary according to the specific situation you're in. The more you understand this, the more effectively you can prepare both your message and your mindset. If you feel confident about what you want to say, it's relatively simple to deliver with energy and impact.

The main points

o   The channel through which you're speaking has an influence over the methods you use, so it's important to take this into account.
o   Your aim with any channel is to understand how it works so you can win the war of attention by removing distractions, speaking with clarity and energy, and generating a human connection with your audience.
o   Online video conferencing is now a common way to present and communicate, with specific rules for doing it well.
o   Other channels include recording to camera, media interviews, podcast interviews and speaking in person.

Chapter 10
Winning the war of attention

It's a busy world out there, and few people are hanging around to discover what you have to say. This is because, important though your expertise may be, and vital though your goals are to you, to your audience you're only one person out of many who are trying to grab a slice of their attention.

This is where the magic of speaking comes in, and this is the whole purpose of what you've been learning in this book. In today's diverse media environment, in which online video reigns supreme, if you can address people so that they listen to and understand your ideas, you've won the first half of the battle. The second half is motivating them to do what you want, which comes from the energy and persuasiveness of how you put yourself across.

Anyone can do this, and so can you. To be a compelling speaker, there are only five things you need to possess:

1. A full appreciation that it's your responsibility to communicate well, not your audience's to understand.
2. Empathy with your audience's mental and emotional state, and with their level of understanding of your topic.
3. Knowledge about the varying characteristics of both in-person and digital environments.
4. Effective ways of structuring and delivering your content for the appropriate audience and channels so that it's clear and motivating.

5. The willingness to prepare and practise so you can speak fluently and handle questions with confidence.

All of this can be learned. Few of us are 'born speakers' – we have to make a conscious decision to commit to becoming good. The rewards are enticing: more advancement of the science and medicine that's so vital in the world today, more awareness of the importance of your work and more influence in your field. People need to know what you know, and they need it now. There's no time to waste in delivering your message, so let today be the first day of your new speaker journey.

Chapter 11
Physical and vocal exercises for speaking success

Your physical state and vocal tone are critical to communicating successfully, whether you're speaking on stage, on camera or virtually. When we look at people, we can usually tell immediately if they're feeling anxious or on edge, so it's important to calm your nerves and prepare your voice by warming up before you speak. Then you can deliver with confidence.

You can see videos of some of these exercises at https://scientificallyspeaking.com/physical-and-vocal-exercises.

Physical exercises

The exercises that follow have the aim of enabling you to tap into your physical power and energy, improving your presence when speaking in public. When you feel grounded and relaxed, it shows in the way you use your body, and this not only helps your audience to trust what you're saying but also enables you to enjoy the experience.

Full body check

Time: two to three minutes

You can practise this exercise anywhere, and the beauty of it is that no one knows what you're up to. Its purpose is not to change anything, but simply to explore what's going on in your body and connect with it. This is important because your posture communicates to others what you're feeling, not just when you're talking but also when you're listening. When you regularly observe your own physical experience, you become more alert to reading the body language of other people, which can contain valuable information.

1. Stand up.
2. Notice your posture. Your impulse will be to change something now you're noticing, but don't. Just be aware of how you stand. Are your legs shoulder-width apart or do you feel less well-balanced than that? Do you feel comfortable? How are your shoulders? Are you pulling them back or slumping forward? Are you standing with your knees locked or are you slightly flexing them?
3. Use active verbs to tell yourself (silently) what you're doing: 'I'm pushing my chin forward; I'm stiffening my back; I'm collapsing my chest; I'm breathing shallowly.' When you use active verbs, you're practising conscious awareness without judgement.
4. Are you holding your jaw tightly or is it relaxed? Notice what's going on with your neck muscles. Are you constricting your throat or are you opening it? Or are you doing something in between? We have a tendency to squeeze ourselves when under pressure.
5. What's going on with your hands? Are you stretching them out or balling them into a fist, or something in between? How much? Is your impulse to cross your arms or let your arms fall by your side?
6. What else are you noticing? Observe how standing in your default way makes you feel. Pay attention and name that feeling.

That's all there is to it. When you make this exercise a habit (and you can do it while waiting for the elevator), you'll realise that your changes in posture are related to how you feel. Noticing is the first step to change, and this exercise gives you a baseline for how you use your body. You have to spot where you hold tension before you can let it go.

Micro-flex

Time: one to three minutes

This exercise helps you to relax. You can do much of it sitting down if necessary, but stand up if you can. Because it involves micro movements that no one can see, you can practise it anytime and anywhere.

1. Flex your toes a couple of times. Unlock your knees slightly. Flex your hands, one at a time. Do the same with your fingers. Lift your shoulders and let them slowly sag.
2. Clench and unclench your buttocks, which is where we tend to store tension.
3. Flex and contract your brow, temples and other muscles of the face. Pay particular attention when you slightly clench and unclench your jaw (be gentle with your neck and throat). Notice the effect this has on your breathing.
4. Go through your body, minimally flexing and contracting your muscles. What other muscles are recruited when you clench and unclench? Can you relax them a little? Don't overdo it – less is more.

Move it

Time: one to two minutes

This is a classic actor's preparation for performance, and you'll want privacy for it. It involves making expansive, relaxed movements that help you to deepen your breath and let go of tension. Often when we're stressed or nervous, we contract our bodies, trying to take up a smaller amount of space. This exercise helps to counteract this tendency as it helps you relax and take up more room.

1. Stand up with feet hip-width apart.
2. Unlock your knees. Flex your hands and feet one at a time.
3. Draw circular motions with each foot in turn. Shake out one leg and then the other.
4. Do the same for your arms and hands. Let your shoulders droop. Make some shoulder rolls.
5. For the next part, ensure there's space around you. Put your hands on your hips and rotate them as if you were hula-hooping. Swing your arms

from side to side. Think 'relax' – yawning is good. Make circles with your arms. Take up lots of space and be as big as you can.
6. Notice how you felt before you did this exercise and how you feel now you've done it.

Power posing

Time: one to two minutes

This is a warm-up exercise you can do before going on stage, and it will help you to feel confident and assured. Social psychologist Amy Cuddy claims that expansive, open postures affect people's feelings, behaviour and even hormone levels, and that this can make us feel more powerful (something we could all do with before a stressful speaking event). Try 'power posing' yourself and see what it does for you. It's best if you practise doing it before your speaking event as well as on the day itself, in which case you'll need somewhere private (if you're at the venue, the bathroom is the best place). You only need a couple of minutes but spend more time if you can.

1. Stand up straight with your feet hip-width apart.
2. Relax your breathing by taking some deep breaths.
3. Make yourself as big you can. You can put your hands on your hips comic-book hero style or splay them out above your head as though you've just won a gold medal in the Olympics. Holding this position can take some effort in the beginning, so only do it as long as you can without tensing your neck muscles. With practice, you'll be able to hold it for longer.

Make some faces

Time: one to two minutes

These exercises bring an expressive aliveness and warmth to your face. If you were seven years old, you'd be a natural at making faces, but as a normally inhibited adult you'll want privacy. A bathroom cubicle or your car will suffice.

1. Scrunch up your face as if you've just tasted something bitter. Then relax.
2. Make an extreme smile. Then relax.

3. The big surprise: describe an 'O' with your mouth, lift your eyebrows, and widen your eyes. Then relax.
4. Puff out your cheeks. Then relax.
5. Move your jaw from side to side, then back and forth. Then relax.
6. Stick your tongue out and open your mouth wide. Describe circles with your tongue. Then relax.
7. This next exercise will also help you reduce tension in the mouth and jaw area. Put the heels of each hand below your cheekbones. Relax your jaw and gently massage it by making circular motions with your hands.

Breathing

Time: five minutes

Breathing has a direct impact on your posture and on how you feel. It also affects the quality of your voice, which has a significant influence on how you're perceived. Learning to do it correctly takes effort and practice because you're working muscles that have been under-used. However, when you get the hang of it you'll find that it takes less effort to breathe and you'll also breathe more slowly, which has a calming effect. Babies and small children naturally breathe well, but around the time of adolescence many of us begin to physically restrict ourselves. Unless trained, the majority of adults don't breathe properly.

As an experiment, take a deep breath. Notice whether your shoulders rise and your stomach draws in. If they do, you're not breathing correctly and you're likely to run out of breath when you speak. This will cause you to trail off at the end of utterances and will reduce your ability to project your voice.

Your diaphragm is a large, dome-shaped muscle underneath your lungs. When you inhale, your diaphragm pushes down, allowing the lungs to expand and fill with air. When you exhale, your diaphragm reverts back to its dome shape, pushing air out of the lungs. You can locate your diaphragm by putting your fingers just beneath your rib cage. When you inhale correctly, you should notice an expansion in the lower rib cage and abdomen.

Experimenting further, put both hands on your lower back on each side of your spine so that you can feel your rib cage. Take a deep breath, and you should feel your hands move apart as your diaphragm pushes down to make room for the inhaled air. When you breathe out, feel your hands move back to their original position.

The following exercise is a good way to learn to breathe correctly.

1. Find a comfortable spot to lie on the floor, and put a pillow beneath your knees and another under your head. Place one hand on your chest and the other just below your rib cage.
2. Breathe slowly through your nose and feel your hand on your stomach rise as you inhale.
3. As you exhale, purse your lips to create some resistance and tighten your stomach muscles as you push out the last remaining air. You can add sound if you like by making a 'ssss' as you exhale.
4. Notice what's happening to the hand on your chest – it shouldn't move much at all, as the movement should be in your stomach.

I recommend that you practise this exercise three or four times a day at the beginning, then you can try doing it from a sitting position and later from a standing position. Once you have the hang of it you can do it anywhere for a few minutes.

Vocal exercises

When you're speaking, you are your instrument; your voice carries emotion in a way no other part of you can. Vocal exercises help you to project your voice and speak with clarity so you're understood. You'll want to select a private place to do these exercises, but first let's take a look at how your voice works.

Your larynx, or voice box, is located between the base of your tongue and the top of your trachea, or windpipe. Two bands of smooth, mucous membranes called vocal folds are stretched horizontally across the larynx. When you're not speaking, the vocal folds are open, thereby allowing air into the lungs. When talking, the folds snap together as you exhale, causing them to vibrate. The quality of your voice depends on the shape and size of your vocal folds.

Because your vocal tissue is made of mucous membranes, it needs to be kept warm and moist. The act of speaking dries out your vocal folds, so make sure you sip enough fluids in the run-up to your talk. Also avoid clearing your throat, shouting and coughing, as these can cause damage to your voice.

Lip buzz

Time: one minute

Put your lips loosely together and blow through them so they vibrate. Vary the pitch of the sound from low to high. Yes, it does look and sound funny, but it loosens tension in your lips and raises your energy. This helps with vocal projection and clarity as it moves your voice forward towards your audience.

Work your tongue

Time: one to two minutes

Giving your mouth a workout is important because its job is to enunciate your words clearly. This exercise loosens your tongue, and engages your vocal cords and breath.

1. Imagine you're brushing your teeth with your tongue. Let it explore all of your teeth.
2. Roll your 'r' sounds: put your tongue behind your upper teeth, exhale and trill your tongue making the sounds. Try varying the pitch and be gentle – don't force anything.

Hum

Time: three to five minutes or longer

Studies have shown that humming is good for your health, and many people claim that doing it for five minutes has a calming effect, which is no small benefit for someone about to speak in public. Some people sing in the shower (a steamy environment is ideal for vocal exercises) and others do it to warm up their voice because humming helps you loosen up tight vocal cords. You can use humming to warm up and to cool down again after you've taxed your voice – for instance, after a long period of talking.

1. Inhale. On a slow out-breath, say the word 'hum' with the emphasis on a long 'm' sound as if you were exaggerating a sigh. Feel the vibration in your forehead, jaw and lips (also known as your vocal mask). If you don't feel the vibrations immediately just try to relax more. Give it time. Play around with your hum.

2. Next, hum with a long nasal 'n' sound. Do this with a big smile and your tongue behind your upper front teeth. You should now experience a different vibration in the sinuses. Find what is comfortable for you by humming from high to low. Don't force anything and be gentle; your humming doesn't have to be loud.

Woo it

Time: one minute

This is a vocal-cord stretch exercise as it loosens and lengthens your vocal cords. Make an 'o' with your mouth as if you're drinking through a straw. Take a long in-breath, and as you exhale through your imaginary straw make a vibrating 'woo' sound.

Enunciate

Time: two to four minutes

This exercise helps you to enunciate your syllables clearly, thereby avoiding a common problem with speaking, which is when you trail off at the end of your sentences. It also gives your vocal apparatus a workout, and regular practice will strengthen the muscles in your face and mouth. It's particularly useful for fast speakers, because learning to enunciate well slows down your speech pattern and makes it easier for your listeners to understand you.

When performing enunciation exercises, exaggerate the shapes your mouth and lips make – imagine someone is trying to read your lips. Open your mouth wider than you usually would and show your teeth.

1. Read the following rhymes aloud a few times:

   Oh what a to-do to die today at a minute or two 'til two
   A thing distinctly hard to say yet harder still to do
   For they'll beat a tattoo at twenty to two
   With a rattatta tattatta tattatta too
   And the dragon will come when he hears the drum
   At a minute or two 'til two today
   At a minute or two 'til two
   Unique New York, New York unique
   She says she shall sew a sheet

2. And try repeating these vocalisations:

   Pa Ta Ka Pah
   Pa Ta Ka Paw
   Pa Ta Ka Poo
   Pa Ta Ka Pee
   Pa Ta Ka Pay
   La La La
   LaLa LaLa LaLa
   LaLaLa LaLaLa LaLaLa
   Ba Da Ga Bah
   Ba Da Ga Boo
   Ba Da Ga Bee
   Ba Da Ga Baw
   Ba Da Ga Bay

Read aloud

Time: three minutes or longer

Reading aloud improves your speech rhythms, enunciation, awareness of intonation and verbal fluency. You can use it as part of your warm-up or as a separate practice technique.

Consider this phrase: 'She is a thief.' Try saying it three times, accentuating a different word each time.

   She is a thief.
   She *is* a thief.
   She is a *thief*.

The words are the same but the intention (communicated by the stress word) is different. The first sentence identifies who the thief is – it's her and not the person over there. The second sentence refutes a claim that she isn't a thief. The third sentence defines what she is – a thief and not, say, a bus driver.

Choose something easy to begin with; nursery rhymes can give you a good sense of speech rhythm. Pay attention to how the words sound because your rhythms and vocal variety make a significant difference to how much people pay attention to what you're saying. Read a paragraph and mark where you want to stress words, and where you'll pause for effect. Record yourself and see how you improve, moving on to longer and more challenging passages as you progress.

Sing in the car

Time: as long as you want

There's not much more to say except to start by gently humming, sing loudly, don't strain your voice and enjoy.

The author

Jo Filshie Browning is a specialist in verbal communication skills. She helps scientists and physicians to speak with impact and authority. She's passionate about conveying the value that science and medicine bring to the world, whether it be by bringing life to data, raising awareness of rare diseases, illustrating the potential of new technologies or driving best practice in patient care. Her purpose is to elevate the communication of science and medicine to positively impact society.

Jo is the founder and Managing Director of Filshie Browning Associates, and its Principal Consultant. She is also the CEO and founder of Scientifically, a training company that supports doctors and scientists to communicate science. In all, she's had 25 years of experience helping scientists and physicians to improve their content, competence and confidence, so they can speak with impact and authority. This enables them to influence the audiences they want to reach and to enhance their reputations.

Jo began her working life at BAFTA and Emmy award-winning Insight News Television, where she trained as a journalist. She then moved into public relations and has held senior positions at GSK and also major international communications agencies in London and Australia, where she worked as a media trainer. Her English degree from Oxford University, her Diploma in Public Relations and her Postgraduate Diploma in Science and Society all underpin her work. She has also lectured undergraduates on science and corporate communications, and held the position of Senior Lecturer at the University of Rhine Waal in Germany.

Today, Jo coaches and trains scientific and medical professionals to become influential speakers and presenters, and her knowledge of media training, including presenting in person and on camera, is in demand across all scientific sectors. Through exercises and role-play, her clients improve their ability to communicate their ideas and build their confidence. Overall, she's trained more than 10,000 people across 43 countries and four continents.

If you'd like to talk to Jo about speaking opportunities, or speaker training and coaching, please visit www.scientificallyspeaking.com.

Acknowledgements

I have had the privilege of working with extraordinary scientists throughout my career and I always learn something from everyone I work with. I'm so grateful to the researchers who shared their science with me and in particular those who took the time to help me understand which parts of the training and the work that we had done together had been the most valuable.

I am also grateful to Professor Alexander Gerber and the University of Rhein Waal who first gave me the opportunity to develop my work formally into a series of lectures – which ultimately set me on the path to this book.

It takes a team to bring a book to the world and I'd like to thank Ginny Carter for the extraordinary professional guidance throughout the writing process. Anthony Lewis took my terrible scribble drawings and created the wonderful graphics to illustrate the book. Alison Jones and the amazing team at Practical Inspiration Publishing were fantastic for diligently taking the book through to publication.

On a personal level, I am particularly grateful to a number of friends and colleagues in encouraging me to start the work, persevere with it and finally to publish it. I couldn't have done this without Ruth Slater, Catherine Minogue, Euan Turner, Juliet Morley and Michelle Grabham who have been sounding boards, critics, proof-readers and cheerleaders. Your support was wonderful.

Finally, I would like to acknowledge with gratitude the support and love of my family. Firstly, to my parents, Marcus and Sheila, and to Angus and Danny for always believing in me and encouraging me. And most importantly to my husband Jason and my children Kit and Will for their support and patience with me during the process of writing; especially during the evenings and days of holiday the book absorbed on top of the challenges of 2020 and lockdowns.

Thank you all!